Reconstruction of China's Low-Carbon City Evaluation Indicator System

A Methodological Guide for Applications

Reconstruction of China's Low-Carbon City Evaluation Indicator System

A Methodological Guide for Applications

Pan Jiahua
Zhuang Guiyang
Zhu Shouxian
Zhang Ying

Chinese Academy of Social Sciences, China

社会科学文献出版社
SOCIAL SCIENCES ACADEMIC PRESS (CHINA)

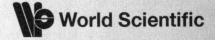
World Scientific

Published by

World Scientific Publishing Co. Pte. Ltd.

5 Toh Tuck Link, Singapore 596224

USA office: 27 Warren Street, Suite 401-402, Hackensack, NJ 07601

UK office: 57 Shelton Street, Covent Garden, London WC2H 9HE

Library of Congress Cataloging-in-Publication Data
Pan, Jiahua, 1957–
    Reconstruction of China's low-carbon city evaluation indicator system : a methodological guide for applications / Pan Jiahua, Zhuang Guiyang, Zhu Shouxian, Zhang Ying.
        pages cm
    Includes bibliographical references.
    ISBN 978-9814612838
    1. Sustainable urban development--China. 2. Carbon dioxide mitigation--China. 3. Urban ecology (Sociology)--China. 4. City planning--Environmental aspects--China. 5. Environmental policy--China. 6. Urban policy--Environmental aspects--China. I. Title.
    HT243.C6P36 2015
    307.1'4160951--dc23
                                        2014038402

British Library Cataloguing-in-Publication Data
A catalogue record for this book is available from the British Library.

重构中国低碳城市评价指标体系:方法学研究与应用指南。

Originally published in Chinese by Social Sciences Academic Press
Copyright © Social Sciences Academic Press, 2013

In-house Editor: Lum Pui Yee

Typeset by Stallion Press
Email: enquiries@stallionpress.com

Printed in Singapore

Preface

Since September 14, 1950, when Switzerland established diplomatic relations with China, there have been frequent high-level visits between the two governments. Trade and economic cooperation increased, and collaboration in various fields including education, tourism, training, science and technology was strengthened. Switzerland has been a pioneer in terms of cooperation between China and European countries. It was one of the first few Western countries to establish diplomatic relations with China. Upon China's reform and opening-up, Swiss companies took the lead in entering China's market and establishing joint ventures. Switzerland was also the first continental European country to recognize China's full market economy status. All these were the result of the pioneering and innovative spirit that has been embedded in Sino-Swiss bilateral relations.

Climate change has become the common challenge for human society and a widespread concern around the world. In response to this challenge, Switzerland and China signed a memorandum of cooperation on climate change. China has achieved remarkable results for low-carbon development as it has been vigorously promoting ecological civilization and actively responding to climate change. Switzerland appreciates China's progressive climate change policies, broad actions and various achievements and is willing to work with China to further cooperate on policies and pragmatic projects. Such cooperation in the field of climate change will be the new highlight in the two countries' bilateral relations, and will contribute to the efforts of addressing global climate change challenges.

In June 2010, the Swiss Agency for Development and Cooperation launched the Low-Carbon City in China project jointly with its Chinese partners. The project aims to build the capacity for urban planning and municipal management, developing a vision, strategy, action plan and supervision system to promote low-carbon city development, enabling best practice exchange and knowledge sharing, developing comprehensive yet China-specific low-carbon standards and management systems, and promoting the development of renewable energy, green building and sustainable transportation. Since the launch of the project, seven cities (Yinchuan, Dezhou, Meishan, Dongcheng District in Beijing, Baoding, Kunming and Jianchuan) have joined the project in the past three years, giving a strong impetus to China's low-carbon city development.

The Institute for Urban and Environmental Studies at the Chinese Academy of Social Sciences is a leading institute for low-carbon economy research and low-carbon city development consulting. It is funded by the Swiss Agency for Development and Cooperation as the consultant to develop China's Low-Carbon City Evaluation Indicator System. Such research aims to study and learn from the European Energy Award and the Swiss Energy City Project and apply their successful experience to guiding China's low-carbon city construction with consideration of China's actualities. We are pleased to see that, through three years of in-depth research, the project team has developed a low-carbon city evaluation indicator system for China, and constructed a framework and application guidelines to evaluate low-carbon cities with methodological support, and successfully tested them in some of the aforementioned pilot.

Right at the time when China's Prime Minister Li Keqiang is visiting Switzerland, the deliverable report of this research project has become ready for publication. I would like to express my sincere appreciation for the hard work of the project team led by Mr. Pan Jiahua, director of the Institute for Urban and Environmental Studies at the Chinese Academy of Social Sciences, as well as warm congratulations on their achievements in this project. At present, the Chinese government has started two batches of national low-carbon provinces and pilot low-carbon cities, followed by numerous other regions and cities that are also endeavoring to transition themselves into low-carbon society. I believe that the publication of the research results will help provide guidance to China's low-carbon city development practice.

Dr. Philippe Zahner
Counselor of the Embassy
of Switzerland in China
Swiss Agency for Development
and Cooperation

June 5, 2013

Foreword

The development of a low-carbon economy requires both theoretical guidance and practical solutions. Cities are the major agents in national economic development; hence national goals of low-carbon economy development will eventually need to break down to the city level and cannot be achieved without concrete implementation of low-carbon construction in the cities. With the domestic low-carbon pilot provinces and cities booming, more and more cities have begun to put forward low-carbon development plans and policies based on their own characteristics. However, with all the different understandings of the concept of a low-carbon city and unguided practice, there is an urgent need to develop a comprehensive evaluation indicator system.

Since 2008, the Institute for Urban and Environmental Studies at the Chinese Academy of Social Sciences has taken the lead in China's low-carbon city evaluation indication system research. Building on its definition of the concept of a low-carbon economy, the institute constructed an indicator system consisting of four clusters of indicators: low-carbon output, low-carbon consumption, low-carbon resources and low-carbon policies. The system has been tested in several cases including the low-carbon roadmap development for Guangyuan City, Jilin City, Shenzhen City and Huangshi City. The media has reported this indicator system as "China's low-carbon city evaluation criteria".

As an important component of the cooperation between China and Switzerland in the field of climate change, the Swiss Agency for Development and Cooperation in China launched the Low-Carbon City in China project in 2010 and entrusted the Institute for Urban and Environmental Studies at the Chinese Academy of Social Sciences to learn from international experience and further develop China's Low-Carbon City Evaluation Indicator System in order to provide standards and guidance for China's low-carbon urban construction. After nearly three years of research, the institute developed a low-carbon city evaluation framework with methodological support for China and application guidelines.

This indicator system is a tool developed with funding from the Sino-Swiss low-carbon city project with the target of guiding China's low-carbon city development. It was developed by drawing on the successful experience of the European Energy Award and Swiss Energy City Project with adjustments based on China's specific situation. The indicator system has been tested in several cities, including Dezhou City in Shandong, Baoding in Hebei, Kunming in Yunnan, Meishan in Sichuan, Yinchuan in Ningxia, and Dongcheng District in Beijing.

China's Low-Carbon City Evaluation Indicator System consists of three integral and inter-connected parts: a list of indicators, a low-carbon city assessment report and action plans for the case study cities. The core of this system, the list of indicators, is further divided into two parts: major indicators and supporting indicators. The evaluation reports are the result of applying the indicator system as well as a review and summary of the pilot cities' low-carbon construction, while the action plans provide recommendation for future improvement.

The theoretical basis of the methodologies for China's Low-Carbon City Evaluation Indicator System is the well-known PDCA in management science. Through the continuous "plan–do–check–act" cycle, experience can be summarized and lessons can be learned for future improvement. The most important feature of this indicator system is that it is not only able to evaluate the status of low-carbon city development, but it is also able to measure the efforts of the cities while taking into account their different geographical features and natural resource endowment. The indicator system could help cities to understand their current achievements, identify their strengths and weaknesses and learn from the successful experience of other cities, so that they can advance their low-carbon transition process in a more efficient way.

As this is exploratory and cutting-edge research, throughout this project there were several adjustments to research objectives. During the study, Mr. Walter Meyer, former Counselor at the Swiss Agency for Development and Cooperation, was extensively involved in the project and provided detailed technical guidance. Moreover, Dr. Wang Liyan, Mr. Liu Ke, Dr. Bai Chenxi, Miss Li Yang and Miss Bai Jie also gave great support for the project implementation. In particular we would like to thank Mr. Philippe Zahner, the incumbent counselor, who always found constructive solutions whenever the project encountered difficulties and was stalled. He played a critical role in helping the project to finally succeed.

The support given by the project management office is highly appreciated. Mr. Zhang Ruijie, director of PMO, together with project managers Mr. Jin Qing, Mr. Weng Weili, Mr. Xie Hongxing and Mr. Wang Boyong are professionals with both expertise and enthusiasm for the research project. They coordinated and participated in all field trip studies of this project.

We also acknowledge the support provided by government officials in the pilot cities, including those from Dezhou, Baoding, Kunming, Meishan, Yinchuan and Dongcheng District in Beijing. Under the coordination of PMO, the research team conducted several field trips in these pilot cities for interviews, indicator testing and data collection. Special thanks to Dezhou Energy Monitoring Commission, the School of Economics and Management at North China Electric Power University, and the carbon indicator system research team at Kunming Low-Carbon Development Research Center.

We owe the success of the project also to Mr. Robert Harbaty, a Swiss expert who brought in advanced concepts and methodologies from Europe, and to Mr. Jiang Zhaoli, Division Chief in Department of Climate Change in National Development and Reform Commission, who provided both guidance and motivation for the project team.

Dr. Pan Jiahua, director-general of the Institute for Urban and Environmental Studies at the Chinese Academy of Social Sciences, was the leader of the research team. Dr. Zhuang Guiyang (Senior Research Fellow) was the acting team leader, and Dr. Zhu Shouxian

(Associate Research Fellow) took responsibility for overall coordination of the project. Dr. Zhang Ying (Associate Research Fellow), Dr. Xiong Na, Dr. Cui Yuqing, Dr. Yuan Lu, Dr. Zhou Yamin, Mr. Li Qing (Senior Engineer), Dr. Mu Haoming and Mr. Liang Benfan participated in the research and made important contributions to the project.

Dr. Pan Jiahua and Dr. Zhuang Guiyang have presented the research output on a variety of occasions, including both domestic and international conferences. During the United Nations Climate Change Conference in Durban in December 2011, the project team presented the research output at the China Session and the exhibit booth of the China Sustainable Development Research Center at the Chinese Academy of Social Sciences, attracting wide attention and interest among conference participants. We are pleased to see that the results of three years of hard work have finally met with readers, yet more importantly, we also hope to see more and more cities benefiting from the application of such an indicator system and guidelines in their low-carbon development. We are willing to provide further intellectual support for cities' low-carbon development efforts, as always.

Professor Pan Jiahua
Director-General of Institute for Urban and
Environmental Studies
Chinese Academy of Social Sciences
May 30, 2013

Project Team

Consultants

Philippe Zahner, Counselor, Swiss Agency for Development and Cooperation

Walter Meyer, former Counselor, Swiss Agency for Development and Cooperation

Jiang Zhaoli, Division Chief, Department of Climate Change at the National Development and Reform Commission

Research Team

Team leader:

Pan Jiahua, Professor and Director-General, Institute for Urban and Environmental Studies, CASS

Executive team leader:

Zhuang Guiyang, Professor, Institute for Urban and Environmental Studies, CASS

Leading Researchers:

Zhu Shouxian, Associate Research Fellow, Institute for Urban and Environmental Studies, CASS

Zhang Ying, Associate Research Fellow, Institute for Urban and Environmental Studies, CASS

Cui Yuqing, International Cooperation Center, Ministry of Environmental Protection

Xiong Na, Post-Doctoral Fellow, Institute for Urban and Environmental Studies, CASS

Yuan Lu, PhD student, Institute for Urban and Environmental Studies, CASS

Li Qing, Senior Engineer, Institute for Urban and Environmental Studies, CASS

Other Participants:

Liang Benfan, Research Fellow, Institute for Urban and Environmental Studies, CASS

Zhou Yamin, Post-Doctoral Fellow, Institute for Urban and Environmental Studies, CASS

Mu Haoming, Post-Doctoral Fellow, Institute for Urban and Environmental Studies, CASS

Support Team from Kunming

Zheng Yixin, Vice President, Senior Engineer, Kunming Institute of Environmental Science

Zhang Dawei, Associate Director, Engineer, Kunming Institute of Environmental Science

Li Zhongjie, Engineer, Kunming Institute of Environmental Science

Fu Rong, Assistant Engineer, Yunnan Chengshui Environmental Engineering Co. Ltd

Xu Yilei, Assistant Engineer, Yunnan Chengshui Environmental Engineering Co. Ltd

Yang Zhi, Engineer, Kunming Institute of Environmental Science

Mi Wei, Assistant Engineer, Yunnan Chengshui Environmental Engineering Co. Ltd

Support Team from Baoding

Wang Jingmin, Professor and PhD Advisor, Department of Economics and Management at North China Electric Power University; Research Fellow, Baoding Low-Carbon Development Research Institute

Yin Xulong, Senior Engineer, Baoding Low-Carbon Office

Wu Wenxin, Senior Engineer, Baoding Low-Carbon Office

Ge Xiaolei, Senior Engineer, Baoding Low-Carbon Office

Sun Wei, Lecturer, Department of Economics and Management at North China Electric Power University; Researcher, Baoding Low-Carbon Development Research Institute

Zhang Shuguo, Associate Professor, Department of Economics and Management at North China Electric Power University; Research Fellow, Baoding Low-Carbon Development Research Institute

Kang Junjie, PhD candidate, Department of Economics and Management at North China Electric Power University

Zhu Daoping, PhD candidate, Department of Economics and Management at North China Electric Power University

Support Team from Dezhou

Wang Shiyan, Senior Economist, Party member of Dezhou Economic and Information Technology Commission; Director, Dezhou Energy Monitoring Commission

Yin Hongkun, Senior Engineer, Deputy Director, Dezhou Energy Monitoring Commission

Ren Xianggui, Senior Engineer, Dezhou Energy Monitoring Commission

Yu Lei, Dezhou Energy Monitoring Commission

Wang Guodong, Engineer, Dezhou Energy Monitoring Commission

Xin Ting, Dezhou Energy Monitoring Commission

Project Management Team

Zhang Ruijie, Director, Sino-Swiss Low-Carbon City Project Office

Jin Qing, Administrative and Financial Manager, Sino-Swiss Low-Carbon City Project Office

Weng Weili, Technical Manager, Sino-Swiss Low-Carbon City Project Office

Wang Boyong, Low-Carbon Transportation Manager, Sino-Swiss Partnership Low-Carbon City Project Office, Sino-Swiss Low-Carbon City Project Office

Xie Hongxing, Low-Carbon Industry Manager, Sino-Swiss Low-Carbon City Project Office

Contents

Chapter One

Introduction

1.1 Background

The development of a low-carbon economy requires both theoretical guidance and practical solutions. To complete the transition from a fossil-based carbon-intensive economy to a low-carbon one requires thorough research on the concept, the implications and the roadmaps of a low-carbon economy. The wide-spread low-carbon transition efforts in China also necessitate a comprehensive indicator system to evaluate and guide such practice.

Cities are the major agents in national economic development; hence national goals of low-carbon economy development will eventually need to break down to the city level and cannot be achieved without forceful implementation of the low-carbon schemes of the cities. With the domestic low-carbon development going deeper, more and more cities have begun to put forward low-carbon development plans and policies in accordance with their own situation. However, no consensus has been reached on how to define the concept of a low-carbon city, while the policies and initiatives are ad hoc, sporadic as well as uncoordinated. No well-designed or coordinated system has been developed yet.

Against this background, it is imperative that a comprehensive evaluation of the current low-carbon development of Chinese cities be undertaken. As the building of low-carbon cities concerns multiple aspects of city development, it requires a full-blown theory, a set of evaluative methods as well as criteria as the bases for such an evaluation without any doubt. Ideally, such an evaluative system should

1

serve multiple purposes. It should not only reflect the current state of low-carbon development of Chinese cities, it should also take local conditions as well as resource endowment into consideration with an eye toward the contribution to low-carbon development from each city. In the meantime, such a system should focus on how to help each city realize the current state as well as overcoming the deficiencies in low-carbon development. In doing so, the advantages and the disadvantages of each city can be detected and successful experience in low-carbon development of some cities can be shared. More importantly, this can promote scientific transition to the low-carbon development of Chinese cities.

Since 2008, the Institute for Urban and Environmental Studies at the CASS has spearheaded China's low-carbon city indicator system research. Having defined the concept of a low-carbon economy, the Institute constructed an indicator system consisting of four dimensions: low-carbon output, low-carbon consumption, low-carbon resources and low-carbon policies. Such a system has been applied to research on the roadmap of low-carbon development of Guangyuan, Jilin, Shenzhen and Huangshi. And this system is believed to be the criteria for evaluating cities' low-carbon development from the media's perspective.

In the absence of a statistical, monitoring and management system for carbon emissions in China (most cities do not even have their own energy balance sheet), the Chinese Academy of Social Sciences established a low-carbon city evaluation indicator system from the macro level, enabling cities to compare and adjust their low-carbon city planning and construction. However, with the progress of low-carbon practices, the evaluation system and methodologies also need to be updated to match the new conditions — the selection of indicators need to better reflect the essence of cities' low-carbon actions, and some qualitative indicators with regard to policies need to be improved.

As an important component of the cooperation between China and Switzerland in the field of climate change, the Swiss Agency for Development and Cooperation in China launched the Low-Carbon City in China project in 2010 and commissioned the Institute for

Urban and Environmental Studies at the CASS to develop China's Low-Carbon City Evaluation Indicator System by referring to the successful experience of the international community in order to provide standards and guidance for China's low-carbon urban construction. After nearly three years of research, the Institute developed a low-carbon city evaluation framework by referring to the experience of European Energy Award (EEA).

1.2 Literature Review

It is critical to have a low-carbon economy evaluation indicator system to guide China's low-carbon city construction. With the rapid development of low-carbon city building, research on a low-carbon indicator system also became very popular. Different indicator systems are developed and released by distinct entities, including local governments, research institutions, universities, industries and non-governmental organizations. These research results are presented mostly in three ways. First, the papers published by academic journals focusing on the discussion of principles of establishing an evaluation indicator system, the basis of selecting indicators, methodologies to non-dimensionalize indicators and weight-setting, combined with conducting specific urban case studies. These studies are primarily academic. Second, some local governments introduced a low-carbon economy indicator system for their region. Such an indicator system is more action-oriented and region-specific, but less systematic or applicable to other regions due to lack of adequate theoretical support. Third, academic groups have been publishing ecological/green/low-carbon/sustainable evaluation frameworks. Such frameworks have larger research scope and more influence, yet they are not focused on carbon reduction.

Beijing Technology and Business University developed the Suining City Regional Green Economy Indicators (2010) using the analytic hierarchy process. This indicator system consists of 70 indicators in 6 categories: resource, consumption, environment, society, economy and governance. These indicators, all quantitative ones, are divided into reference indicators, guiding indicators and binding indicators. As the system has many indicators and all of them are quantitative ones, it is difficult to apply it to a large number of cases. Columbia University, Tsinghua University and McKinsey & Company jointly established the China Urban Sustainability Index (2010) aimed at assessing cities' overall sustainability from five dimensions: basic requirements, adequacy of resources, environmental health, the built environment and commitment to sustainability. This index consisted of 18 indicators and was relatively simple to apply considering data availability. However, the report did not explain in-depth the basis or the logic of selecting such indicators.

The Economist Intelligence Unit (EIU) developed in collaboration with Siemens the European Green City Index (2010), which mainly measures the current environmental performance and commitments to reducing the future environmental impact of major European cities. The European Green City Index consists of 30 indicators in 8 categories: carbon emissions, energy, building, transportation, water, waste, land use, air quality and environmental governance. Out of the 30 indicators 16 are quantitative ones designed to measure cities' current conditions and 14 indicators are qualitative ones aimed at evaluating their willingness or ambition.

The European Energy Award® established a set of indicators for standardized assessments. Its most prominent feature is that it is an action-oriented system, hoping to improve energy-related urban activities through a gradual process. The European Energy Award assesses six dimensions: development and regional planning, urban buildings and facilities, supply and disposal, transportation, internal organization, communication and cooperation. The system has a total of 118 indicators: 27 major indicators and 91 sub-indicators. Different kinds of European Energy Awards are given to cities according to their implementation of relevant measures guided by this system. Cities will receive the European Energy Award if they implement 50% of the measures, or the European Energy Award — Gold if they implement 75%. Currently more than 600 cities in 9 countries have participated in this program.

The World Eminence Chinese Business Association (2010) initiated the Green City contest for China's mainland cities, aimed at discovering best practices, showing the cities' economic growth and harmonious development of resources and the ecological environment, and assisting China's green economy development. The Green City award evaluates candidate cities for their green economy (resource consumption and environment degradation caused by urban economic development), their green economic potential (the city's economic health and sustainable development potential), their human settlements environment (the level of comfort citizens enjoy) and their pollution prevention (awareness and action for urban environmental pollution prevention).

The Chinese Society for Urban Studies (2011) introduced the Urban Ecological Livable Development Index (UELDI), which adopted a vector structure assessment methodology to assess cities' ecological construction from the soft aspect (behavioral process) and the hard aspect (results), then position the cities in different quadrants accordingly in order to seek their reasonable development path. The UELDI positions cities into four quadrants: the starting phase, developing phase, naturally green phase and enhancing phase. The higher value a city gets, the more livable it is. The goal of this index is to guide cities' development by the public's values and standards.

In conclusion, most low-carbon city evaluation systems focus on assessing cities' status quo while paying a certain level of attention to policies and efforts. It is impossible to conduct a simple comparison among all indicator systems as they each have a different scope, different focus and different definitions of the concepts.

1.3 Logical Framework for the LCCC Indicator System

The LCCC indicator system is the research output of the Sino-Swiss Low-Carbon City in China project. It was developed by both drawing on the experiences learned from the European Energy Award and considering China's specific national conditions.

The LCCC indicator system has three mutually supporting parts: the list of indicators, low-carbon city assessment reports and action plans for the cities evaluated. The list of indicators is divided into the

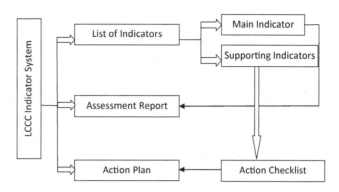

Figure 1.1   Logical Framework for the LCCC Indicator System.

primary indicators and supporting indicators. It is the core of the indicator system. The assessment report is the result of applying the indicator system and presents a review of these cities' low-carbon initiatives. Action plans provide recommendations and guidance for future improvement (see Figure 1.1).

The list of indicators has two parts: primary indicators (15 indicators in 5 categories: low-carbon economy, low-carbon energy, low-carbon infrastructure, low-carbon environment and low-carbon society) and supporting indicators (52 indicators in 4 categories: urban management, green economy, green building and low-carbon transportation). These two parts are closely connected yet with different emphases. Primary indicators evaluate cities' low-carbon development status and their efforts, while supporting indicators focus on guiding cities' future actions.

The LCCC primary indicator system is an external evaluation system, so it needs to be simple and widely applicable. Supporting indicators provide in-depth assessment of the cities and so are more action-oriented — they can guide cities to learn from best practices and develop their own action plans. These two parts are closely connected, as the former yields evaluation results and the latter can be used as a tool to guide actions. On the other side, results can be achieved through different ways. For example, in order to improve energy intensity, cities can choose to increase the energy efficiency of different sectors — including industry, transportation, building or others. Therefore, the categories of primary indicators and supporting indicators are not the same, while the indicators in the two systems are not identical either.

Application of the LCCC indicator system will yield two main outputs: the Low-Carbon Development Assessment Report and the Recommended Actions for Low-Carbon Development. The assessment report gives a comprehensive assessment of the low-carbon city development level and analyzes its development prospects. An Action Plan is drawn up by applying the supporting indicator system which identifies gaps, making specific recommendations for action and providing clear guidance for urban development planning. The Low-Carbon Action Plan is derived from the related checklist. It clarifies specific actions in key areas, priority level, responsible person (organization), time schedule and budget.

The theoretical basis for developing the LCCC indicator system is the well-known PDCA concept. Through the continuous PDCA cycle, experience of success or lessons of failure are gradually summarized to guide continuous improvement in the next cycle. Based on this, the LCCC indicator system applied the three-step approach to combine comprehensive indicator assessment with an action plan in a dynamic way in order to increase cities' low-carbon development performance. A similar approach has been used by other projects in Europe (European Energy Award and Swiss Energy City) successfully for over 20 years, where the program has attracted more than 1000 cities.

The three-step approach adopted is as follows:

Step 1: Conduct investigation and research using the LCCC indicator system (both primary and supporting indicators) to assess a city's low-carbon development status, then identify priority areas to improve low-carbon city management using the supporting indicators. Data will be collected through interviews, workshops, public statistics, the official website, statistical yearbooks and newspapers and media reports.

Step 2: Conduct monitoring and evaluation using the LCCC indicator system and application guidelines, then compose assessment reports.

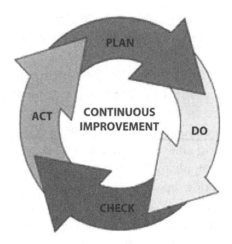

Figure 1.2   Process of Low-Carbon City Development in China.

Step 3: Develop action plans using low-carbon action plan templates. Assessment results will be integrated into these action plans.

1.4 Primary Indicators in the LCCC Indicator System

Primary indicators are the critical tool in the LCCC indicator system as well as the basic criteria for low-carbon development evaluation. Representative indicators are chosen to enable a comprehensive and systematic assessment of a city's low-carbon development. Primary indicators evaluate both a city's absolute achievements as well as its efforts through comparative studies with its peer cities and the average. This system could be used both for a city's overall assessment and for comparison among different cities.

Primary indicators can support action-oriented energy and climate policies for Chinese cities, with a focus implementation of national and provincial programs at the city level. The system was developed with consideration for variety and can accommodate differences among cities in terms of size, levels of economic development and carbon reduction potential. It can guide cities to develop their Five-Year Plan and set their low-carbon development goals (including both qualitative and quantitative ones). It also helps defines the "low-carbon city" concept and enables "low-carbon city development" to be measurable, reportable and verifiable.

1.4.1 *Theoretical Basis*

At present, low-carbon city evaluation methodologies are rather dispersed and there is no systematic theory yet developed. The theoretical basis of developing a low-carbon city evaluation indicator system includes low-carbon economic theory, sustainable development theory, Environmental Kuznets curve hypothesis, decoupling theory, the ecological footprint theory, etc.

The low-carbon economy is a concept arising in the context of climate change. A low-carbon economy refers to an economic pattern which is achieved at a given level of carbon productivity and human development, with certain carbon emission constraint, aiming to achieve the Global Shared Vision of greenhouse gas emission control. A low-carbon economy can be achieved through technological leaps and institutional regulations. It is shown through energy efficiency increase, energy structure optimization and rational consumer behaviors. The low-carbon economy concept has three core characteristics, namely low-carbon emission, high carbon productivity and phases. To evaluate whether a society has transitioned to a low-carbon economy or not, we must consider four essential elements: its resource endowment, technological progress, consumption patterns and stage of development.

The concept of sustainable development was first proposed in Stockholm in 1972 at the United Nations Conference on the Human Environment. In 1987, the World Commission on Environment and Development published "Our Common Future", which defined sustainable development as "development that meets the needs of the present without compromising the ability of future generations to meet their own needs". Sustainable development emphasizes the coordinated development of social, economic, cultural, resources, environment and other aspects of life. It requires the vector composed by indicators in these areas to be monotonically increasing (strong sustainability), or at least that on the whole the trend is not monotonically decreasing (weak sustainability). The three pillars of sustainable development are economic development, social development and environmental protection.

Decoupling refers to achieving economic development while reducing resource consumption and negative environmental impact. If fossil fuel consumption and carbon emission shows a very small positive growth relative to economic growth or urban development, it is called relative decoupling, while if it shows zero or negative growth, it is called absolute decoupling. The key to low-carbon economy development is to achieve the decoupling of economic growth and energy consumption/carbon emissions, with the latter achieving low/zero or even negative growth. Decoupling theory proved that low-carbon development is possible to be achieved, but it cannot be achieved without policy intervention tools such as planning, standards, pricing and public awareness management.

An ecological footprint refers to the area of biologically productive space required per person in order to maintain the person's current lifestyle through the provision of resources and eco-services. The ecological footprint calculates resources consumed by each person with a standardized productive geographical area, and by calculating the total supply and total demand of the ecological footprint — ecological deficit or ecological surplus — it becomes possible to assess different regions' contribution to the global ecological state. Similarly, the carbon footprint concept is also used to calculate greenhouse gas emissions in order to measure human behavior impacts on the surrounding environment. The carbon footprint usually is measured by both direct emissions and indirect emissions.

1.4.2 Design Principles

An indicator system should have the appropriate academic support and integrity. As the primary indicator system for evaluating low-carbon cities involves various activities and different resource endowments, the development of such a system should follow several principles as explained below.

Scientific and systematic design: The indicator system should be designed to fully reflect all low-carbon city connotations. It should be able to assess both achievements and efforts. Meanwhile, the indicator system is not a single concept or standard, as it must be able to fully and accurately reflect all aspects of low-carbon governance. It should be neither complicated with too many indicators nor oversimplified with too few indicators.

Holistic and hierarchical approach: The indicator system should be designed with different hierarchical layers, while all indicators should focus on the core concept of low-carbon development. This way, the system can avoid blind spots or overlaps, so it can reflect both overall performance and detailed facts in different aspects. A properly weighted design is also important to get a balanced picture.

Comparability and applicability: The indicator system should have both a relative evaluation system and an absolute evaluation system, taking into account the universality of indicators and the variety of evaluation objects so it can reflect each city's characteristics as well as overall performance. All indicators selected are quantitative ones to ensure that the evaluation results can be compared over time and among different cities.

1.4.3 *Evaluation System*

The LCCC primary indicator system is built by using the analytic hierarchy process, which is based on the aforementioned principles and theoretical basis. The system can be categorized into five groups, including: low-carbon economy, low-carbon energy, low-carbon infrastructure, low-carbon environment and low-carbon society. This system consists of key indicators reflecting the three pillars of sustainable development: economic prosperity, environmental protection and social equity. It also includes major contents of low-carbon development such as green building, low-carbon transportation, industry decarbonization and energy decarbonization.

The system consists of 15 indicators (see Table 1.1) in 5 categories: (1) economic transformation indicators; (2) social transformation indicators; (3) infrastructure decarbonization indicators; (4) resource indicators; and (5) environmental indicators. Among them, the economic indicators measure a region's low-carbon economic development stage and overall performance of the region in

Table 1.1   LCCC Indicator System: Primary Indicators.

Category	No	Indicator	Rationality of Selecting This Indicator
Economy	(1)	Carbon productivity	The 12th Five-Year Plan set carbon intensity reduction targets, which are decomposed at the city level for local implementation. Carbon intensity is the inverse of carbon productivity, which reflects a city's carbon competitiveness.
	(2)	Energy intensity	The 11th Five-Year Plan and 12th Five-Year Plan set emission reduction targets, which are decomposed at the city level for local implementation. Energy intensity reflects the efficiency of an economy.
	(3)	Decoupling index	The level of economic development and the overall carbon emissions show a certain relation. The decoupling index reflects carbon emission characteristics in different development stages.
Energy	(4)	Percentage of non-fossil energy in primary energy consumption	The 12th Five-Year Plan set explicit targets for the non-fossil ratio in primary energy consumption targets. Although this indicator is subject to regional characteristics, it could still guide a city's low-carbon development.
	(5)	Per capita non-commercial renewable energy use	The energy mix cannot be easily adjusted at the city level. Utilization of non-commercial energy sources such as solar water heaters, small biogas, geothermal, PV, etc., could still assist low-carbon development.
	(6)	Carbon intensity of energy	Upgrading the energy mix and increasing the supply of clean energy are critical for low-carbon transition. This indicator evaluates a city's energy mix.

(Continued)

Table 1.1 (*Continued*)

Category	No	Indicator	Rationality of Selecting This Indicator
Infrastructure	(7)	Energy consumption per unit building area for public buildings	Green building is an important area for low-carbon transition, and government agencies should set an example to improve building energy efficiency in public institutions.
	(8)	Energy consumption per building area for residential buildings	Residential building energy efficiency is closely related to citizens' quality of life. This indicator reflects both residential energy efficiency and energy consumption behaviors.
	(9)	Ratio of green transport	Green transport is another important area for low-carbon transition. This indicator reflects the public transport status and transport infrastructure conditions.
Environment	(10)	Percentage of days with API less than 100	Low-carbon city development helps reduce urban diseases and can bring synergies. This indicator measures the urban environmental quality.
	(11)	Domestic water consumption per capita per day	Resource consumption and energy consumption are closely related. This indicator reflects resource consumption patterns of citizens.
	(12)	Forest coverage rate	This indicator reflects a city's carbon sink capacity of its potential for climate change adaptation.
Society	(13)	Urban-rural income ratio	The 12th Five-Year Plan emphasizes inclusive growth. This indicator reflects social equity and the gap between urban and rural residential energy consumption.
	(14)	Per capita CO_2 emission	The per capita carbon emission level reflects a city's level of low-carbon development.
	(15)	Low-carbon management institution	This indicator reflects the level of carbon management and policy efforts.

implementing national energy reduction targets; low-carbon energy indicators evaluate the city's low-carbon energy resources as well as its effort for energy structure optimization; infrastructure decarbonization indicators assess carbon content in urban infrastructures and low-carbon consumption levels in the region. Low-carbon environment indicators measure the city's green development; and low-carbon society indicators evaluate the society's consumption patterns and social equity.

Selection of the 15 indicators was based on 5 criteria: relevance to low-carbon, content diversity, their specific characteristics, policy-oriented and accommodating regional differences.

In terms of content diversity and specific characteristics, although carbon productivity is closely related to other indicators — decoupling index, energy intensity and carbon intensity of energy — each of these indicators has different implications. Carbon productivity is the converse of carbon intensity, which is linked with carbon footprint and can reflect a city's competitiveness. Carbon intensity indicators have become national binding targets. With regard to indicator diversity, each indicator has a different focus. Carbon productivity measures efficiency, the decoupling index reflects the relation between carbon emission and economic development at different stages, while energy carbon intensity evaluates energy mix.

In terms of policy orientation and low-carbon relevance, indicators on the air pollution index (API) and domestic water consumption are selected for the primary indicators. They seem to have weak links with low-carbon development, but as a matter of fact, water supply and consumption have a strong link with energy as many processes require heating and pumping, while the air pollution index is selected to show the close relation between human health/life quality and low-carbon development. Environment status is a reflection of low-carbon development and environmental improvements need to be achieved through low-carbon development. The adoption of these environmental indicators aimed to stress that the goal of low-carbon development is to enhance instead of sacrificing the quality of life.

In terms of accommodating regional differences, indicators are chosen to reflect carbon competitiveness like carbon productivity, as well as carbon emission characteristics during different stages of development such as decoupling index. In the low-carbon energy indicators an important environmental indicator proposed by the Twelfth Five-Year Plan, namely the ratio of non-fossil fuels in primary energy consumption, was also considered. However, this ratio can be largely different among different regions, while the role of the city government to influence this ratio is rather limited. Therefore, we introduced the indicator of per capita non-commercial renewable energy usage, which can reflect the local government's efforts. In addition, per capita carbon emission is selected as an indicator for low-carbon society, as consumption patterns are becoming increasingly relevant to China's low-carbon development, although no relevant assessment criteria is set at the national level yet.

1.4.4 *Evaluation Criteria*

The LCCC primary indicator system is built to evaluate both a city's low-carbon construction status and its level of effort for comparison. As cities can be in different stages of development and have different resource endowments, it would be controversial to simply compare their low-carbon development status. The highlight of the LCCC primary indicators is that it does not only objectively describe a city's low-carbon development status, but also evaluates the city's effort for low-carbon transition.

Building assessment indicators require weight allocation, which is very important for multi-factor-based decision-making. Most weight allocation systems adopt a subjective weighting method where weight is assigned based on researchers' judgment. The guiding ideology of the LCCC primary indicator system weights assignment is to highlight the core indicators' importance to reflect the city's various aspects of low-carbon development and its transformation features as well as requirements of related national policies. At the present stage no weights are set for the indicators, but rather, each indicator is assessed respectively. Considering China's Five-Year Plan planning system, assessment conducted currently is mainly based on results of the Eleventh Five-Year Plan period.

We take carbon productivity as an example for quantitative indicators assessment. Carbon productivity and carbon intensity are reciprocal. During the Eleventh Five-Year period neither the state nor local government has set targets to reduce carbon intensity, but carbon intensity can be derived through processing publicly available data. Therefore, the indicator's assessment will be done through comparing the city's carbon intensity and national carbon intensity during the Eleventh Five-Year Plan period. If by the end of the Eleventh Five-Year Plan period the carbon intensity growth rate of a city reaches 80% of the national level, then 60% is assigned; similarly, 80% is assigned if it reaches 100% of the national level, and 100% is assigned if it reaches 120% of the national level.

From the methodology theory perspective, evaluation and indicator comparison is based on ladder comparison, i.e. cities are divided into groups relative to the reference (e.g. of 60%, 70%, 80% of the national level). However, this method is not fair for cities around the critical points. As a result, actual assessment is based on linear comparison. Current assessment focuses only on fossil-energy-related carbon emission, without considering the other sources of greenhouse gases, because data is relatively easy to collect for the former while it accounts for the majority of total GHG emissions.

For qualitative indicators we take carbon management institution as an example. Carbon management institution evaluates a city's low-carbon development management system through its management structure, responsibilities and implementation mechanisms. This indicator analyzes the city government's political will and institutional setup for carbon reduction. A city gets 20% of the score if it has established a low-carbon development work group, then another 10% if the mayor/municipal party committee secretary participates in the work group, then another 20% if there is clear distribution of responsibilities and authority, then another 20% if the city has developed a low-carbon city/economy development plan, then another 20% if climate change adaptation strategies are composed, then another 10% if it has established a working group for an energy management system (EMS).

1.5 Supporting Indicators in the LCCC Indicator System

The LCCC primary indicator system is matched with the supporting indicators which can assist with in-depth assessment of the city. They are also very action-oriented as they can guide cities to adopt best practices and develop low-carbon action plans. Supporting indicators focus on assessment of cities' concrete actions and implementation. Their main audience is city managers and policy-makers. Its main object is to evaluate and guide city administrators to assess their management system rather than technical details. The results of such evaluation are the basis to develop the city's low-carbon action plans.

1.5.1 *The Four Categories*

Supporting indicators cover four main areas, including urban management, green economy, green building and low-carbon transport (Table 1.2). This work defines a "city" as the administrative scope of one municipality, including both its urban and rural areas. City administrators can advance low-carbon development through efforts and actions in these four areas. Urban economic indicators reflect the city's development stage and its industry characteristics. The economy is the city's main subject of low-carbon transition. Through proper urban planning and management, cities can reduce the risk of adverse effects of climate change and transition to a low-carbon economy. Buildings and transport are closely related to people's lives and are two areas with the fastest growth of carbon emissions that need special attention.

(1) Urban Management

As a country's basic administrative unit, cities are responsible for both the country and their industries and citizens. Cities play an important role in addressing the global challenge of climate change. However, irrational urbanization and development patterns make cities in developing countries more vulnerable to climate change. Besides, the urban environmental infrastructure and disaster prevention ability still need to match the pace of economic development in many cities. The critical role cities play in global greenhouse gas emissions determines their huge responsibility in addressing climate change. It is imperative to tackle climate change through strong involvement of the cities in the areas of capital, technology, policy and social awareness. Reality shows that proper urban planning and management can help reduce the risk of adverse effects of climate change and assist the city to transition to a low-carbon economy. A city's competitiveness in the next few decades as a friendly platform for business and residents depends largely on its ability to adjust in the low-carbon economy era.

(2) Green Economy

Cities as spatial carriers of productivity bring together various resources such as capital, labor, science and technology, and serve as the network of economic entities, the hub for economic growth, and the platform for trade and innovation. Judging from the economic structure of China's provincial capital cities and major municipalities since 2008, we find that the proportion of secondary industry is still high in most cities. Most large and medium-sized cities have a very low proportion of primary industry, generally accounting for 10% or less, except for a few cities such as Nanning, Harbin, Chongqing, Shijiazhuang and Fuzhou. The total proportion of secondary industry and tertiary industry generally looks the same, but two thirds of cities have a larger secondary industry than tertiary industry. Industry is important both for the urban economy and low-carbon transition. By promoting industrial energy efficiency, increasing the use of renewable energy, developing the circular economy, green industry and environmental industry, etc., cities can greatly reduce their carbon emissions. The primary industry and service industry are also important for low-carbon transition; the former is more important in underdeveloped regions while the latter plays a larger role in developed regions.

(3) Green Building

Green building has important implications for low-carbon development and sustainable urbanization. As China continues its rapid urbanization, its total building energy consumption growth remains at a high rate and shows a trend to go even higher. According to the Ministry of Housing, Urban and Rural Development (MOHURD), currently national building energy consumption accounts for about 28% of total end-use energy consumption. In accordance with the experience of developed countries, this proportion will gradually increase further to 30–40%. If energy consumption for building materials production and building construction are accounted for, the proportion will exceed 40% at present. Studies and practice showed that building energy saving has the largest potential and is the most direct and effective pathway to energy conservation. With the guidance of the MOHURD, all cities are working on promoting green building through regulations, policies, technical standards and market mechanisms. Low-carbon buildings can be achieved mainly by improving building energy standards and ensuring that new buildings comply with such standards, retrofitting existing buildings, strengthening energy-saving operations and management for construction, and promoting the applications of new forms of energy in green buildings.

(4) Low-Carbon Transport

China's transport accounts for more than 20% of total energy consumption. It relies heavily on oil and gas, accounting for about 40% of petrol consumption, about 95% of gasoline, 60% of diesel and 80% of kerosene. Cities in China usually have high density, intensive land use and high population density. The high-density urban space layout enables intensive use of urban land — it reduces travel distance for residents and encourages non-motorized travel modes. However, in recent years China's private car ownership and car travel has kept rising while the bicycle travel rate has declined at an average annual rate of 2–5%. The reason why more and more people prefer to travel by car is largely because urban planning and transport infrastructure development is often car-oriented. China's urban transportation infrastructure has long lagged behind — there was limited urban road construction and the lack of high-grade urban roads convinced people of the need for large-scale and high-intensity urban road construction. However, policy-makers failed to acknowledge the need for public transport development and public-transport-friendly land use planning. Meanwhile, land leased became a major source of city government revenue, and municipalities' urge to sell land, coupled with unrealistically high population projections, led to irrational expansion that also resulted in increases in residents' travel distance. On a more micro level, urban planning and design emphasize motorized traffic facilities but ignore slow traffic systems and mixed land use, resulting in uneven distribution of road space where walking and cycling roads are severely squeezed and traffic safety cannot be guaranteed.

1.5.2. Evaluation System

The evaluation system is designed in three aspects: planning, implementation and management. The goal of supporting indicators and action checklists is to help raise awareness comprehensively and systematically among urban administrators, helping them identify gaps of low-carbon management and improve their work.

1.5.3 Action Checklist

Supporting indicators are a combination of both quantitative and qualitative ones. A total of 52 indicators were selected, each with definition, scope, importance and data source considered. Each indicator can help develop an action checklist, which consists of controlled items, general items and optional advanced items classified according to the importance of each item. Controlled items are the necessary actions to be taken to achieve low-carbon development and are the basis for other actions. General items are measures to be taken to further development. Optional advanced items are those actions needed in order to achieve an advanced level of low-carbon development and require more effort. An action checklist is developed in accordance with the role, responsibility and authority of a specific administrative department, focusing on its carbon management system rather than technical details.

By comparing reality to the checklist, supporting indicators can help concerned departments in city governments to set their own

Table 1.2 Supporting Indicators in LCCC Indicator System.

Indicators

Urban management

1.1 Planning

    1.1.1 Low-carbon inventory tool implemented

    1.1.2 Low-carbon development strategy established

    1.1.3 Integration of low-carbon concepts into urban planning

    1.1.4 Ratio of municipal funding for low-carbon activities (renewable energies, energy saving and environmental conservation) in local government budget

1.2 Implementation of low-carbon planning and policies

    1.2.1 Incentives for low-carbon and green economy promotion

    1.2.2 Number of demonstration projects of low-carbon communities/ schools/hospitals/supermarkets, etc.

    1.2.3 Green procurement (e.g. labeled electrical appliances — 1 or 2 class energy label)

    1.2.4 Information disclosure/accessibility for low-carbon planning and management documents and data

1.3 Low-carbon management for utilities

    1.3.1 Energy consumption for district heating

    1.3.2 Coverage rate of urban population with access to gas (%)

    1.3.3 Share of local renewable energy production in total energy consumption (%)

    1.3.4 Energy consumption/m^3 water supplied

    1.3.5 Energy consumption for sewage water treatment

    1.3.6 Water conservation measures

    1.3.7 Treatment rate of municipal solid waste

    1.3.8 Municipal solid waste sorting and reduction

Green economy

2.1 Low-carbon industries

    2.1.1 Non-CO_2 GHG emission in industries and reduction measures

    2.1.2 Implementation of investment policies for energy-intensive and highly polluting industries

    2.1.3 Elimiating obsolete capacity

    2.1.4 Energy efficiency compliance rate for industrial enterprises above designated size (%)

    2.1.5 EMS application rate among priority enterprises for energy efficiency (%)

    2.1.6 Number of priority enterprises for energy efficiency that have signed voluntary agreements for emission reduction

    2.1.7 Ratio of renewable energy/energy efficient appliances product value in total industrial outputs

    2.1.8 Application of renewable energy in industrial sector

    2.1.9 Recycling and utilization rate of industrial solid waste (%)

(Continued)

Table 1.2   (*Continued*)

Indicators

2.2 Low-carbon services

 2.2.1. Number of trained energy managers in hospitals, schools, large shopping malls, hotels, airports, etc.

 2.2.2 Ratio of major service industries that signed voluntary agreements for emission reduction

 2.2.3 Number of companies in energy service, energy management and contracting, CDM services, energy consulting, etc.

2.3 Agriculture

 2.3.1 Fertilizer consumption (ton/ha)

 2.3.2 Share of rural households with biogas digesters in households suitable for building biogas digesters (%)

 2.3.3 Treatment and reuse of agriculture and forestry residues

 2.3.4 Energy-saving measures applied in agricultural production

Green building

3.1 Planning

 3.1.1 Application of energy consumption statistical tool for different building types

 3.1.2 Green building action plan developed

 3.1.3 Building energy consumption planning of new urban area developed

3.2 Green building management

 3.2.1 Overall performance of building energy management

 3.2.2 Efficiency improvement of existing buildings

 3.2.3 Application of renewable energies in buildings

 3.2.4 Capacity building for green building

 3.2.5 Promoting green building (narrow definition)

 3.2.6 Implementation of incentive and policy instruments

 3.2.7 Demonstration projects of best available low-carbon technologies

Low-carbon transport

4.1 Low-carbon transport strategy and planning

 4.1.1 Emission inventory for different transportation modes established

 4.1.2 Low-carbon transportation strategy and action plan developed

4.2 Transport management

 4.2.1 Integrated transport management in place

 4.2.2 Average commuting time

 4.2.3 Share of new energy vehicle in public vehicles (%)

(*Continued*)

Table 1.2 (*Continued*)

Indicators

4.2.4 Measures to promote efficient/renewable energy-efficient bulbs for street lighting

4.2.5 Assessment of public transport service

4.2.6 Number of related awareness-raising events per year

4.2.7 Number of persons trained in low-carbon capacity-building activities per year

4.2.8 Planning and maintenance of slow traffic infrastructure

low-carbon goals, to measure the gap and identify opportunities for improvement, to develop a new implementation plan and launch new low-carbon activities.

In the checklist items for action are classified according to their importance and are specifically described. Judgment is easy to make based on its "Yes/No" design which greatly simplifies the application. For some quantitative indicators, there will be a list of data described in details, and such data source is generally from this same department involved.

Considering the cities' geographical constraints or different levels of development, a third choice of "Not Applicable" is added besides "Yes" and "No" for cases where they cannot take certain specific actions due to objective constraints or relevant data is not yet available. In this case, reasons why this item is infeasible need to be explained. As each city has its own characteristics and demands, it is allowed for a city to add more indicators on top of the existing supporting indicators according to the basis and principles of selection, the set categories and the defined scope.

The LCCC supporting indicators and action checklists can be used to guide cities to improve their work. Taking into account the city's different geographical differences and self-assessment needs, they can make adjustments based on the current version. At present we focus on guiding cities to compare and assess their work with the checklist. Although many supporting indicators are difficult to quantify at this stage, as low-carbon work proceeds it should become possible in the future.

Table 1.3   An Example of an Action Checklist (1.3.1 district heating energy).

Item	Yes	No	N/A
Controlled Items			
Establish specific institution to be in charge of district heating			
Measures implemented to enforce energy regulations in district heating system			
Management scheme on district heating established			
General Items			
Targets are set for energy efficiency improvement			
Implementation plan to achieve the targets is developed			
Establish online monitoring system for heat supply			
Optional Advanced Items			
Set up detailed targets on energy consumption reduction for different phases including generation, transmission and heat loss.			

1.6 Outputs

The main outputs of the LCCC supporting indicator system consist of a Low-Carbon Development Comprehensive Assessment Report and a Low-Carbon Action Plan.

The Low-Carbon Development Comprehensive Assessment Report presents a comprehensive assessment of a city's low-carbon level, its current achievements and future prospects. The report consists of assessment through both primary indicators and supporting indicators. The content of the report includes the following:

— Description of a city's low-carbon development current status and level of effort, mainly based on analysis results with primary indicators;

— Description of a city's low-carbon development current status in four aspects: urban management, economy, green building and low-carbon transport, mainly based on analysis results with primary supporting indicators;

— Low-carbon transition background, practice and highlights;
— Challenges cities face in their low-carbon transition;
— Next-step actions and policy recommendations.

The assessment report is primarily for a single city, focusing on evaluating the development and changes in the city itself, and reflecting its characteristics and highlights on the basis of a comprehensive assessment. Such reports do not aim at comparing different cities (see Table 1.4).

The LCCC indicator system will also provide a list of measures to help cities develop action plans. Preparation of the action plan is based on the evaluation report. The action plan includes a comprehensive list of future actions along with detailed actions, priority, responsible institutions, timetable and budget.

Unlike the action checklist derived from supporting indicators, the Low-Carbon Action Plan provides specific recommendations

Table 1.4 Template of LCCC Assessment Report.

    I. Facts of the City

   II. Backgrounds

  III. Practice and Highlights

  IV. Status Quo and Efforts
     1. Economy
     2. Energy
     3. Infrastructure
     4. Environment
     5. Society

    V. Low-Carbon city Construction and Management
     1. Urban management
     2. Green economy
     3. Green building
     4. Low-carbon transport

  VI. Challenges

 VII. Conclusion and Recommendations

References

Annex: Recommended Low-Carbon Action Plan

List of Team Members

Table 1.5 Template of Low-Carbon Action Plan.

#	Recommended Actions	Responsible Dept.	Timetable	Budget	Priority
Category 1 Urban management					
1					
2					
3					
Category 2 Green economy					
1					
2					
3					
Category 3 Green building					
1					
2					
3					
Category 4 Low-carbon transport					
1					
2					
3					

based on its characteristics, resource endowments, current status and inadequacies, with a clear purpose and operability. The Action Plan shows that the indicators are action-oriented, and can be utilized and incorporated into urban development planning.

1.7 Summary

China has set low-carbon green development as a clear goal in its Twelfth Five-Year Plan. Faced with the dual challenges of a domestic and global green revolution, the country needs to seize opportunities for low-carbon economic transformation. According to the Twelfth Five-Year Plan, China's cities must fulfill their share of responsibilities for energy conservation and carbon emissions intensity reduction allocated by the central and provincial governments.

The LCCC indicator system learns from the success of international projects and considers China's national conditions. It is designed to systematically assess a city's low-carbon management and help it develop low-carbon development plans and an action checklist. It has the following four characteristics:

1. The system consists of two parallel sets of indicators — primary indicators and supporting indicators. Primary indicators summarize a city's low-carbon development and point out the direction that supporting indicators should aim at, while supporting indicators explain how to improve the work evaluated by primary indicators and how to guide urban management practice. The city's efforts and improvement guided by supporting indicators can also be reflected in the assessment by primary indicators. The two have a dynamic, interactive and mutually supporting relationship.
2. Based on an evaluation by both sets of indicators, a low-carbon action plan can be developed to provide specific guidance with proposed measures and necessary implementation tools, and ultimately help the city achieve its goals for low-carbon development.
3. The design of the LCCC indicator system was based on the PDCA (plan, do, check and act) concept and with an action-oriented focus. In addition to external evaluation, the city can apply this tool for self-assessment (the LCCC project team can provide consulting services for this).
4. The LCCC indicator system can help cities to establish clear goals and tasks, to strengthen their supervision, monitoring and evaluation system, and ultimately to achieve the required energy and carbon emission targets and other carbon-related development visions. This would enable the cities to build their competitiveness in the global low-carbon transformation.

Since its inception in August 2010, China's LCCC indicator system project was developed over two years and went through several rounds of testing and debugging in Dezhou, Kunming, Baoding, Meishan, Beijing Dongcheng District and Yinchuan, and has basically achieved the expected goals. We hope the LCCC indicator system can be applied to all pilot cities in the LCCC project as well as other cities across China as soon as possible.

References

1. The Economist Intelligence Unit. 2009. European Green City Index: Assessing the environmental impact of Europe's major cities. A research project conducted by the Economist Intelligence Unit, sponsored by Siemens. Munich, Germany.
2. The Economist Intelligence Unit. 2011. Asian Green City Index: Assessing the environmental performance of Asia's major cities. A research project conducted by the Economist Intelligence Unit, sponsored by Siemens. Munich, Germany.
3. D. Hoornweg, L. Sugar and C. Lorena Trejos Gómez. 2011. Cities and greenhouse gas emissions: Moving forward. *Environment and Urbanization*, 23(1): 207–227.
4. IPCC. 2007. *Climate Change Mitigation, 2007.* UK: Cambridge University Press.
5. Columbia University, Tsinghua University, McKinsey & Company. 2010. *Urban Sustainability Index: A New Tool to Measure Chinese Cities.*
6. Ji Zhu, He Yan, Sun Jin. 2010. *Research on Regional Green Economy Index System in Suining City, Sichuan.* World Economy Research Center in Beijing Technology and Business University, Suining Municipal Development and Reform Commission, Suining City Green Economy Research Institute.
7. World Eminence Chinese Business Association. 2010. *Methodology and Evaluation System for China Green City Award.* Available at http://www.Chinacity.org.cn/csph/pingjia/55691.html (accessed 25 May 2011).
8. Pan Jiahua, Zhuang Guiyang, Zheng Yan, Zhu Shouxian, Xie Qianyi. 2010. Low-carbon economy concept and core elements. *International Economic Review*, 4: 88–101.
9. Zhuang Guiyang. 2008. Low-carbon economy to lead the world economic development. *World Environment*, 2: 34–36.
10. Zhuang Guiyang, Pan Jiahua, Zhu Shouxian. 2011. Low-carbon economy: Connotation and comprehensive evaluation index system. *Economic Perspectives*, 1: 132–136.
11. Zhu Shouxian, Zhuang Guiyang. 2010. Revitalization of Northeastern China from a low-carbon perspective: Case study on Jilin City. *Resource Science*, 2: 230–234.
12. Lei Hongpeng, Zhuang Guiyang, Zhang Chu. 2010. *China's Urban Low-Carbon Development: Strategies and Methodologies.* Beijing: China Environmental Science Press.

13. Wang Guoqian, Zhuang Guiyang. 2011. Low-carbon economy: Understanding and development patterns. *Learning and Exploration*, 2: 134–138.
14. He Xindong, Zhuang Guiyang. 2011. Practices and reflection on low-carbon city planning in small cities in Western China: Case study on Guangyuan City in Sichuan Province. *China Population, Resources and Environment*, 21(3): 482–485.
15. Ye Qing, Li Fen, Yan Tao. 2012. UELDI Index for dynamic assessment on eco-cities: Chinese Cities Ecological and Livable Development Index Report. *Construction Technology*, 12: 18–24.
16. Zhang Jinhe, Zhang Jie. 2007. China ecological footprint modeling. *Regional Research and Development*, 2: 90–96.
17. Chen Wu, Li Yunfeng. 2010. Discussions on China's sustainable energy development. *Energy Technologies & Economics*, 22(5): 17–23.
18. Chinese Society for Urban Studies. 2011. *China Low-carbon Eco-city Development Report 2011*. Beijing: China Building Industry Press.
19. Zhang Wenxu, Wang Daqing, *et al.* 2011. Initial research on low-carbon city index for reclamation area in Heilongjiang. Journal of Northeast Agricultural University, 9(3): 23–27.
20. Li Zhongmin, Yao Yu, Qing Dongrui. 2010. Decoupling of industry development, GDP growth and carbon emissions. *Statistics and Policy-making*, 11: 108–111.
21. Zhu Zhi, Zhou Shaohua, Yuan Nanyou. 2009. Developing a low-carbon economy to address climate change: Low-carbon economy and evaluation index. *China National Conditions and Strength*, 203(12): 4–6.
22. World Bank. 2011. *World Development Report 2010: Development and Climate Change*. Beijing: Tsinghua University Press.
23. Song Deyong, Lu Zhongbao. 2009. China carbon emission factors of its periodical fluctuations. *China Population Resources and Environment*, 19(3): 18–24.

Chapter Two

Primary Indicators in the LCCC Indicator System and Application Guidelines

2.1 Economy

2.1.1 Carbon Productivity

(1) Definition

Carbon productivity is the inverse of carbon intensity. The indicator characterizes a city's low-carbon competitiveness.

Unit: 10,000 *yuan* GDP/ton CO_2.

(2) Scope

Carbon emission refers to the amount of carbon dioxide emitted within certain geographic boundaries (in this case a municipality) due to the combustion of fossil fuels and the consumption of electricity (emissions relevant to net imports or net exports of electricity should be considered). Due to data unavailability, other GHGs (e.g. methane, N_2O, HFC) specified in the Kyoto Protocol are not considered for the moment.

The boundary of a municipality is the jurisdictional boundary defined by the central government, including both urban and rural areas.

(3) Instrument

Carbon emissions are calculated using two variables: total energy consumption and energy mix. If a city has an energy balance sheet, total carbon emissions can be directly calculated in accordance with the IPCC carbon emission factors. If it does not have any energy balance sheet, calculation needs to be made based on its energy mix.

31

(4) Assessment Criteria

In the Eleventh Five-Year Plan period, the carbon productivity target was not raised by the central or local government. But the corresponding data of the country and cities can be calculated by the above method. Therefore, this indicator will be evaluated by comparing the city's carbon productivity change rate and the country's carbon productivity change rate during the Eleventh Five-Year Plan period. If by the end of the Eleventh Five-Year Plan period, the carbon intensity growth rate of a city reaches 80% of the national level, then 60% is assigned; similarly, 80% is assigned if it reaches 100% of the national level, and 100% is assigned if it reaches 120% of the national level.

2.1.2 *Energy Intensity*

(1) Definition

Energy intensity is also known as the energy consumption per unit of output value. It refers to the energy consumption in an area or region within a certain period of time. The unit is tce/10,000 *yuan*. At the country level, this indicator is calculated with the ratio of total domestic energy consumption or total end-use energy to its GDP.

(2) Scope

Total energy consumption refers to the total consumption of energy of various kinds by the production sectors and the households in a given period of time. It is a comprehensive indicator to show the scale, composition and increase rate of energy consumption. Total energy consumption includes that of coal, crude oil and their products, natural gas and electricity. However, it does not include the consumption of fuel of low calorific value, bio-energy and solar energy. Total energy consumption can be divided into three parts: end-use energy consumption; loss during the process of energy conversion; and energy loss. GDP is calculated in constant value of the currency.

(3) Instrument

$$\text{Energy intensity} = \frac{\text{Total energy consumption}}{\text{GDP of the city}}.$$

(4) Assessment Criteria

The national energy saving target was set during the Eleventh Five-Year Plan period and was allocated to local governments. Therefore, the assessment of this indicator will be given by comparing the city's energy intensity change rate and the country's energy intensity change rate during the Eleventh Five-Year Plan period. If it reaches 80% of the national level, then 60% is assigned; similarly, 80% is assigned if it reaches 100% of the national level, and 100% is assigned if it reaches 120% of the national level.

2.1.3 *Decoupling Index*

(1) Definition

Decoupling refers to producing more economic wealth with less material consumption than the past. This concept is proposed considering that for a long time economic growth is highly dependent on material consumption. Decoupling is designed based on the DPSIR model (Driving forces, Pressures, States, Impacts, Responses) and reflects growth elasticity changes of driving forces (GDP growth for example) and pressures (environment pollution for example). Decoupling can be measured by an index. The concept consists of primary decoupling, secondary decoupling and dual decoupling, or absolute decoupling vs relative decoupling. Primary decoupling refers to the decoupling of economic growth and natural resource use, measured by the relationship between natural resources (energy) consumption and economic growth (GDP); secondary decoupling refers to the decoupling of natural resources and environmental pollution, measured by the relationship between environmental pollution (CO_2) and natural resource (energy) consumption; dual decoupling means both primary decoupling and secondary decoupling are achieved. Tapio (2005) divided decoupling indicators into three categories: coupled, decoupling and negatively decoupling, based on the concept of decoupling elasticity. He then further subdivided them into eight types of decoupling: weak decoupling, strong decoupling, weak negative decoupling, strong negative decoupling, expansion negative decoupling, expansion connection, recession decoupling and recession connection.

(2) Scope

Considering the economy can fluctuate in a short time, this indicator is measured in periods each with five years. The GDP is calculated in constant value of the currency.

(3) Instrument

$$DR_{t0,t1} = \frac{EP_{t1} / EP_{t0}}{DF_{t1} / DF_{t0}}.$$

Where EP_t is the environmental pressure variable at the time of t, DF_t is the economic driving force variable at the time of t.

(4) Assessment Criteria

The assessment criteria is as shown in Figure 2.1 below:

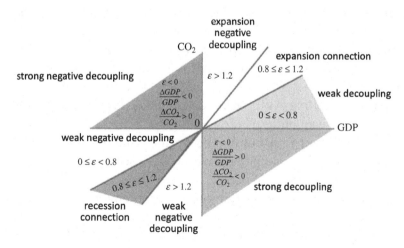

Figure 2.1 Different Types of Decoupling.

2.2 Energy

2.2.1 *Percentage of Non-fossil Energy in Primary Energy Consumption*

(1) Definition

Percentage of non-fossil energy in primary energy consumption (unit: %).

(2) Scope

According to China's Renewable Energy Law, renewable energy refers to non-fossil energy including wind energy, solar energy, hydro energy, biomass energy, geothermal energy, oceanic energy, etc. Primary energy refers to forms of energy that naturally exist in nature, such as coal, crude oil, natural gas, geothermal and direct solar radiation, etc.

Renewable energy includes commercial RE and non-commercial RE. Commercial RE refers to energy integrated in the grid, e.g. most hydro and wind generation. This indicator only considers consumption of to-grid RE.

Total energy consumption refers to the total consumption of energy of various kinds by production sectors and households in a given period of time. It is a comprehensive indicator to show the scale, composition and increase rate of energy consumption. Total energy consumption includes that of coal, crude oil and their products, natural gas and electricity. However, it does not include the consumption of fuel of low calorific value, bio-energy and solar energy. Total energy consumption can be divided into three parts: end-use energy consumption; loss during the process of energy conversion; and energy loss.

(3) Instrument

$$\text{This percentage} = \frac{\text{Non-fossil energy consumption}}{\text{Total Primary energy consumption}}.$$

(4) Assessment Criteria

If a city's ratio in 2010 is higher than the national average, then it gets full score. If it is lower than the national average, then compare its change rate with the national average change rate. If it reaches 60% of the national level, then 40% is assigned; similarly, 60% is assigned if it reaches 80% of the national level, and 80% is assigned if it reaches 100% of the national level.

2.2.2 *Per Capita Non-Commercial Renewable Energy Use*

(1) Definition

This indicator is defined as non-commercial renewable energy consumption per capita.

(2) Scope

Renewable energy refers to non-fossil energy including wind energy, solar energy, hydro energy, biomass energy, geothermal energy, oceanic energy, etc. Consumption of renewable energy includes that of commercial RE and that of non-commercial RE. Commercial energy refers to renewable energy connected to the grid, such as most hydropower and wind generation. Non-commercial energy refers to renewable energy which is not connected to the grid. For example, some small hydropower stations generate power which is only used locally but not connected to the provincial or national grid. Some individual wind farms haven't been connected to the grid yet due to various reasons such as price and transmission barriers. Solar energy is largely used among cities in China, but mostly in the forms of solar heating system, water heating system, street lights, air-conditioning and so on, which is not converted into electricity. This is true with geothermal, tidal or other renewable energies as well. Non-commercial renewable energy in this report refers to solar water heaters, biogas (including those centrally supplied ones), off-grid photovoltaics, geothermal, biomass gasification and biomass carbonization.

(3) Instrument

Currently data is available only for commercial renewable energy consumption, not for non-commercial renewable energy consumption. Therefore, in the initial stages of applying this indicator system, we encourage cities to collect data on non-commercial energy. It is feasible to collect various types of RE consumption data at the city level in China. Solar heating and generation can be converted to tons of coal equivalent based on its heating area, generation efficiency and average radiation hours. Geothermal can be converted to tons of coal equivalent based on its power (kWh). Biomass utilization can also be converted to the equivalent amount of coal based on the amount of biomass used.

(4) Assessment Criteria

Same as above.

2.2.3 *Carbon Intensity of Energy*

(1) Definition

Carbon intensity of energy refers to the emission factor of unit energy consumption (unit: tCO_2/tce).

(2) Scope

Carbon emission refers to the amount of carbon dioxide emitted within certain geographic boundaries (a municipality) due to the combustion of fossil fuels and the consumption of electricity (emissions relevant to net imports or net exports of electricity should be considered). Due to data unavailability, other GHGs (e.g. methane, N_2O, HFC) are not considered for the moment.

Total energy consumption = Initial energy stock + primary energy production + energy import — energy export — final energy stock. Total energy consumption refers to the total consumption of energy of various kinds by the production sectors and the households in a given period of time. It is a comprehensive indicator to show the scale, composition and increasing rate of energy consumption. Total energy consumption includes the consumption of coal, crude oil and their products, natural gas and electricity. However, it does not include the consumption of fuel of low calorific value, bio-energy and solar energy. Total energy consumption can be divided into three parts: end-use energy consumption; loss during the process of energy conversion; and energy loss.

(3) Instrument

$$\text{Carbon intensity of energy} = \frac{\text{Primary energy carbon emission}}{\text{Tons of coal equivalent}}.$$

Information on total energy consumption and energy mix can be found in the national or regional energy balance sheet. If most cities do not have an energy balance sheet yet then estimation based on currently available data is needed.

(4) Assessment Criteria

This indicator assessment only looks into cities' achievement during the Eleventh Five-Year Plan period.

If a city's ratio in 2010 is higher than the national average, then it gets full score. If it is lower than the national average, then compare its change rate with the national average change rate. If it reaches 40% of the national level, then 0% is assigned; similarly, 60% is assigned if it reaches 80% of the national level, and 80% is assigned if it reaches 100% of the national level.

2.3 Infrastructure

2.3.1 *Carbon Emission Per Unit Building Area for Public Building*

(1) Definition

CO_2 emission per square meter of public buildings (unit: $kce/m^2 \cdot year$).

(2) Scope

CO_2 emission refers to the CO_2 emitted by energy consumption in buildings in a statistical year.

The statistical year is a whole year which covers the complete heating season. (Details about the statistical methodology are available in the Statistical Reporting System of Energy Consumption and Energy Conservation Information of Civil Buildings, issued by the MOHURD.)

Public buildings here refer to those in both rural and urban areas, such as office buildings, hotels, restaurants, schools, supermarkets, hospitals, etc. Industrial buildings are excluded.

The energy consumption in buildings includes energy used for HVAC, lighting, elevators, cooking, and electric appliances.

The types of fuel include electricity, gas, district heating and coal. Biomass is excluded due to the lack of data.

Building area refers to the constructed area of a building, i.e. the sum of each floor area to external wall. For details refer to Calculation Rules for Building Area GB/T 50353-2005.

(3) Instrument

$$CO_2 \text{ emission per building area} = \frac{\text{Total } CO_2 \text{ emission}}{\text{Total building area}}.$$

For cities implementing the Statistical Reporting System of Energy Consumption and Energy Conservation Information of Civil Buildings released by the MOHURD in 2010, Table 9 in this document can be used for data calculation. Currently only 79 cities in China have building energy consumption data. Among the LCCC pilot cities, only Yinchuan is among these 79. For other LCCC pilot cities, field research is needed to get data.

Construction area can follow the per capita living area estimation, or follow the heating area estimates. Taking into account the data availability problem, you can count only urban residential building energy consumption per unit area.

According to the Notice on Further Promoting Public Building Energy Efficiency jointly issued by the Ministry of Finance and Ministry of Housing and Urban Construction (MOF-MOHURD [2011] No. 207), China's objective during the Twelfth Five-Year Plan period for public building energy efficiency is to establish and improve the energy monitoring system for public buildings, especially large ones over 20,000 m^2, and to enable monitoring and measuring of public building energy consumption through collecting statistics, energy audits and energy dynamic monitoring and other means. The country also aims to identify public buildings with intensive energy consumption through setting an energy consumption baseline. Moreover, the government will gradually promote retrofitting of public buildings with intensive energy consumption. The overall target set for this is to achieve a 10% reduction in energy consumption per unit area of public buildings, and 15% for large public buildings.

(4) Assessment Criteria

Building carbon emission is a comprehensive indicator reflecting the carbon intensity of buildings. An assessment of this indicator will be made based on data from the Eleventh Five-Year Plan period. Public building carbon emission change rate = Emission of the year/emission of base year.

If the emission change rate is lower than 90%, 100% is assigned; 80% is assigned if the emission change rate is higher than 90% but lower than 100%; 40% is assigned if the emission change rate is higher than 100%.

2.3.2 Carbon Emission Per Building Area for Residential Building

(1) Definition

CO_2 emission per square meter of residential buildings (unit: $kce/m^2 \cdot year$).

Unit: kg standard coal/$m^2 \cdot year$.

(2) Scope

CO_2 emission refers to the CO_2 emitted by energy consumption in buildings in a statistical year.

The statistical year is a whole year which covers the complete heating season. (Details about the statistical methodology are available in the Statistical Reporting System of Energy Consumption and Energy Conservation Information of Civil Buildings, issued by the MOHURD.)

Residential buildings here refer to buildings for residential purposes, such as dormitories, apartments, houses and villas.

The energy consumption in buildings includes energy used for HVAC, lighting, elevators, cooking and electric appliances. The types of fuel include electricity, gas, district heating and coal. Biomass is excluded due to the lack of data.

Building area refers to the area residents live in.

(3) Instrument

$$CO_2 \text{ emission per building area} = \frac{\text{Total } CO_2 \text{ emission}}{\text{Total building area}}.$$

For cities implementing the Statistical Reporting System of Energy Consumption and Energy Conservation Information of Civil Buildings released by the MOHURD in 2010, Table 9 in this document can be used for data calculation. Currently only 79 cities in China have building energy consumption data. Among the LCCC pilot cities, only Yinchuan is among these 79. For other LCCC pilot cities, field research is needed to get data.

Construction area can follow the per capita living area estimation, or follow the heating area estimates. Taking into account the data availability problem, you can count only urban residential building energy consumption per unit area.

(4) Assessment Criteria

The assessment of this indicator will be given based on data from the Eleventh Five-Year Plan period. Residential building carbon emission change rate = Emission of the year/emission of base year.

If the emission change rate is lower than 90%, 100% is assigned; 80% is assigned if the emission change rate is higher than 90% but lower than 100%; 40% is assigned if the emission change rate is higher than 100%.

2.3.3 Number of Public Transport Vehicles Owned Per 10,000 People/Green Transportation Ratio

(1) Definition

This indicator looks into the number of public transport vehicles owned per 10,000 people. The green transportation ratio refers to the ratio of residents choosing public transport or slow traffic system to their total commuting volume. This indicator reflects the development level of public transport and the design of overall transportation. It is an overall and macro concept. Considering data availability, for cities that do not have data on green transportation ratio, only the number of public transport vehicles owned per 10,000 people will be applied.

(2) Scope

Green transportation includes slow traffic modes (walking or cycling) and public transport modes.

Public transport here includes public buses, subways, taxis, ferry, etc. Only the number of bus equivalents is considered. Short-distance shuttle buses are excluded. Rail is not considered either due to lack of data and the fact that it is not easily comparable as not all cities have subways or light rail.

Resident population refers to the de jure population.

(3) Instrument

$$\text{Number of public transport vehicles per capita} = \frac{\text{Number of public buses}}{\text{Resident population}}.$$

(4) Assessment Criteria

If a city's ratio in 2010 is higher than the national average, then it gets full score. If it is lower than the national average, then the ratio it gets is used for assessment.

2.4 Environment

2.4.1 *Percentage of Days with an API of Less Than 100*

(1) Definition

The percentage of number of days with an API lower than 100 in a full year (unit: %).

(2) Scope

The API integrates the concentration of several atmospheric pollutants and calculates them into one index to reflect the level of air pollution. It is convenient to use the API to reflect the air quality of a city over a short period of time. Pollutants considered in the API include sulfur dioxide (SO_2), total suspended particles (TSP), breathable particles (PM10) and nitrogen oxide (NOx). For more details about the API, please refer to Ambient Air Quality Standard [GB3095-1996].

(3) Instrument

Data will be from the *Urban Environment Statistical Bulletin.*

(4) Assessment Criteria

The assessment of this indicator will be given based on comparing the average performance during the Eleventh Five-Year Plan period to base year data.

If the number of blue-sky days is less than 80%, 0% is assigned; 80% is assigned if the number of blue-sky days is more than 90% but less than 90%; 100% is assigned if the number of blue-sky days ≥ 90%.

2.4.2 *Forest Coverage Rate*

(1) Definition

The forest coverage rate refers to the ratio of forest area to total land area in a region, which is often expressed in percentage. It is an important indicator that reflects the status of forest resources and the afforestation level.

(2) Scope

Forest area refers to the area of forest land where trees and bamboo grow with a canopy density above 0.3, including land of natural woods and planted woods, but excluding bush land and thin forest land. It is an important indicator which reflects the total area of forest resources.

Forest refers to tree-dominated land area equal to or greater than $0.0667\,\text{hm}^2$ (1 Chinese acre), with a canopy density equal to or greater than 0.2, in situ growth height of plants equal to or higher than 2 m, including natural forest, artificial young forest, bamboo forest meet this standard, and special shrubbery with two or more lines and 4 m spacing or less, or crown projected shadow of 10 m wide or above.

(3) Instrument

$$\text{Forest coverage rate (\%)} = \frac{\text{Forest area}}{\text{Total land area}} \times 100\%.$$

(4) Assessment Criteria

According to National Forest City standards, the forest coverage rate should reach 35% for southern cities and 25% for northern cities. Cities reaching such standards get full score, otherwise the score is discounted accordingly.

2.4.3 *Domestic Water Consumption Per Capita Per Day*

(1) Definition

Domestic water consumption per capita per day reflects the water consumption status in one region (unit: liter per capita per day).

(2) Scope

Since the data for rural areas is not available at the moment, the indicator will focus on the data in urban areas. Domestic water consumption per capita per day is a commonly used indicator in the statistical yearbook.

Domestic water consumption refers to water supplied by municipal water supply facilities or self-built facilities, and consumed for domestic purposes by urban residents (refer to the Standard for Domestic Water Consumption, GB/T50331-2002).

The population adopted in this indicator refers to the urban de jure population, which is the population who lives in a certain urban area over a certain period of time (usually over six months). For details of this definition refer to Standards for the Sixth National Census.

(3) Instrument

Domestic water consumption per capita per day

$$= \frac{\text{Total Domestic water consumption}}{\text{Residential Population} \times 365 \, (366)}.$$

(4) Assessment Criteria

Domestic water consumption is a useful indicator to reflect the condition of water scarcity in a municipality. Reduction in per capita water consumption will contribute to not only water resource conservation but also energy conservation associated with water supply and waste water treatment, which eventually lead to carbon emission reduction.

Assessment criteria are set as follows: 100% is assigned if it is lower than 80% of the national average; 80% is assigned if it is 100% of the national average; 60% is assigned if it is 120% of the national average; 0% is assigned if it is higher than 200% of the national average.

2.5 Society

2.5.1 *Urban-Rural Income Ratio*

(1) Definition

The urban-rural income ratio is the ratio of urban residents' disposable income per capita and rural net income per capita. This is an important indicator for evaluating development disparity and "the well-off society" proposed by the government.

(2) Scope

Income here refers to the average income of both urban residents' disposable income and rural net income.

(3) Instrument

Urban-rural income ratio

$$= \frac{\text{Urban residents' disposable income per capita}}{\text{Rural net income per capita}}.$$

(4) Assessment Criteria

Assessment criteria are set as below: 100% is assigned if it is lower than 20% of the national average; 80% is assigned if it is lower than the national average; 50% is assigned if it is higher than the national average.

2.5.2 *Per Capita CO_2 Emission*

(1) Definition

Average carbon emission per capita within the jurisdiction area of a municipality in a given year (unit: tCO_2 per capita).

(2) Scope

Carbon emission refers to the amount of carbon dioxide emitted within certain geographic boundaries (a municipality) due to the combustion of fossil fuels and the consumption of electricity (emissions relevant to net imports or net exports of electricity should be considered). Due to data unavailability, other GHGs (e.g. methane, N_2O, HFC) are not considered for the moment.

The population adopted in this indicator refers to the urban de jure population, which is the population who lives in a certain urban area over a certain period of time (usually over six months). For details of this definition refer to Standards for the Sixth National Census.

The boundary of a municipality is the jurisdictional boundary defined by the central government, including both urban and rural areas.

(3) Instrument

$$\text{Per capita carbon emission} = \frac{\text{Total carbon emission}}{\text{Total population}}.$$

(4) Assessment Criteria

The assessment of this indicator will be given based on data from the Eleventh Five-Year Plan period.

China has not set any national target for per capita carbon emission reduction yet, though internationally some other countries have done so. A feasible approach to benchmark cities' per capita emission is to group them based on income. This is a common practice internationally. Municipalities are grouped into two categories: (1) municipalities with per capita income higher than the national average; and (2) municipalities with per capita income lower than the national average. Accordingly, the benchmarking should be set differently for the two categories as well.

With GDP per capita lower than the national average level: 50% is assigned if its carbon emission reaches 120% of the national average; 80% is assigned if its carbon emission reaches 100% of the national average; 100% is assigned if its carbon emission is no higher than 120% of the national average.

With GDP per capita higher than the national average level: 50% is assigned if its carbon emission reaches 100% of the national average; 80% is assigned if its carbon emission reaches 80% of the national average; 100% is assigned if its carbon emission is no higher than 50% of the national average.

2.5.3 *Institution for Low-Carbon Management*

(1) Definition

This indicator evaluates a city's management institution with regard to low-carbon development, including its institutional structure, responsibilities and implementation.

(2) Scope

In general, a low-carbon management leading group and a low-carbon management workgroup are the two critical parts of this institution. The leading group is led by the city mayor and involves directors from all relevant departments or bureaus. They should meet regularly to discuss the low-carbon development. The workgroup reports to the leading group as well as informing the public through websites and other communication tools.

(3) Instrument

Data will be found through local surveys and interviews as well as public records.

(4) Assessment Criteria

This indicator assesses the current status and operation of the low-carbon management institutions. A city gets 20% of the score if it has established a low-carbon development leading group; another 10% if the mayor/municipal party committee secretary is in the leading group; another 30% if it has established a low-carbon management office; another 10% if it has established a workgroup for the energy management system (EMS); another 30% if there is clear distribution of responsibilities and authority.

Chapter Three

Supporting Indicators in the LCCC Indicator System and Application Guidelines

3.1 Municipal Management

3.1.1 *Planning Indicators*

(1) Low-Carbon Toolkit Implemented

Definition

The indicator of the low-carbon inventory tool, typically a spreadsheet or database, is to help the city calculate GHG emissions from different sources. It is helpful for policy-makers to monitor GHG emissions and make informed decisions.

Scope

According to the IPCC's definition, the scope of GHG emissions includes six categories: CO_2, CH_4, N_2O, PFCs, HFCs, SF_6. GHG emissions are mainly from five fields: energy activities, industrial process, agricultural production, land use change and forestry industry, and waste disposal.

Relevance

This indicator is important to evaluate the municipality's effort to calculate the local GHG emission. Municipalities should integrate GHG emission mapping into its planning. The low-carbon inventory could be designed as defined by the IPCC while considering local statistical approaches. In 2010 the NDRC issued the Notice on

Compiling GHG Emission Lists at the Provincial Level (see NDRC-CC[2010]2350).

Instrument

GHG emission = AD (activity data) * EF (emission factor).

Data Sources

Local government work report, DRC, Low-Carbon Workgroup, Economy and Industry Committee.

Data Sheet:

Parameter	Unit	Quantity
The emission of CO_2	10,000 metric tons	
The emission of CH_4	10,000 metric tons	
The emission of N_2O	10,000 metric tons	
The emission of PFCs	10,000 metric tons	
The emission of HFCs	10,000 metric tons	
The emission of SF_6	10,000 metric tons	

Key Responsible Entities for Action

Municipal DRC and Statistical Bureau.

Action Checklist

Item	Yes	No	N/A
Controlled Items			
Is the city listed by the NDRC as a pilot city for implementing a low-carbon strategy?			
Is there any specific administrative office or staff assigned for the low-carbon inventory tool in the municipality?			
Is there any capacity-building plan for a low-carbon inventory tool census?			

(Continued)

(Continued)

Item	Yes	No	N/A
General Items			
Is there any financial support specifically for setting up a low-carbon inventory tool system?			
Does the municipality have a consecutive plan for implementing a low-carbon inventory tool?			
Optional Advanced Items			
Is there any plan to incorporate the indicator of the low-carbon inventory tool into the local statistical yearbook?			

(2) Low-Carbon Development Strategy Established

Definition

This indicator is about how to establish a low-carbon development strategy while taking the development level of the city into consideration. The strategy must be feasible and operable. It is correlated not only to capital and technology, but also to institutions.

Scope

The scope of this indicator includes objectives and quantified targets in terms of low-carbon planning, implementation of low-carbon planning and policies, and low-carbon management for utilities. The scope must be comprehensive, including improving energy efficiency, reducing energy consumption, adopting more advanced technologies and renewable energies.

Relevance

This indicator is important because the rationality of the low-carbon development strategy determines the municipal administration's policy and action. It is also useful to evaluate the efficiency of the low-carbon development strategy and to find room for improvement.

Instrument

Qualitative analysis combined with quantitative analysis.

Data Sources

Municipal DRC, Low-carbon Workgroup, Economy and Industry Committee.

Key Responsible Entity for Action

Municipal DRC.

Action Checklist

Item	Yes	No	N/A
Controlled Items			
Is the low-carbon development strategy incorporated into the city's 12th Five-Year Plan?			
Does the low-carbon development strategy cover energy efficiency, the optimization of industrial structure and carbon sequestration?			
Does the low-carbon development strategy set explicit GHG emission goals, tasks and concrete measures?			
Has the municipality established a supporting policy for the low-carbon development strategy?			
Has the municipality set up a system for GHG emission data statistics and management?			
General Items			
Are there any binding goals and timetable for a carbon emission cap?			
Optional Advanced Items			
Has the low-carbon development strategy set long-term goals, like 2020 or further?			

(3) Integration of Low-Carbon Concepts into Urban Planning

Definition

The indicator of integration of low-carbon concepts into urban planning requires policy-makers to take infrastructures of the gas pipe network and district heating into consideration as a whole. The responsible planning bureau should make an energy development plan and renewable energy plan, binding urban development plan for 2011–2030 and urban resource supply (water, electricity, heat) plan.

Scope

The scope of this indicator covers a broad range of urban energy. This indicator guides cities to conduct wise spatial planning to reduce mobility and energy demands, aim for balanced urban density and pursue maximum use of available waste heat.

Relevance

Integration of low-carbon concepts into urban planning requires scientific thinking for development. Infrastructure development is crucial for a country's economic development and trade practices, but they also need plenty of energy-intensive raw materials. Hence, integration of low-carbon concepts into urban planning is very crucial to future low-carbon development.

Instruments

Qualitative analysis, comparative analysis with historical data.

Data Sources

Municipal DRC, Low-Carbon Workgroup, municipal statistical bureau.

Key Responsible Entity for Action

Municipal DRC, Planning Bureau, Bureau of Housing and Urban-Rural Development, Transportation Bureau, Bureau of Industry and Information Technology.

Action Checklist

Item	Yes	No	N/A
Controlled Items			
Are low-carbon concepts integrated into the local urban planning?			
Are there wise spatial plans enabling saving mobility and energy demands locally?			
Does the "zipper phenomenon" (repeated and low efficiency pipe maintenance work, measured by two or more times of digging ground open per year) frequently appear?			
General Items			
Is renewable energy incorporated into the local energy system?			
Optional Advanced Items			
Does the city have good capacity for prevention and adoption of extreme climate disasters?			

(4) Ratio of Municipal Funding for Low-Carbon Activities (Renewable Energies, Energy Saving and Environmental Conservation) in Local Government Budget

Definition

This indicator is to evaluate the scale of funds spent on low-carbon activities in the local government budget. Low-carbon activities are guided by the concept of sustainable development, aiming at the win-win scenario of economy development and environmental conservation via technology innovation, institutional innovation, industry transformation and new energy development to reduce carbon-intensive materials such as coal and petroleum.

Scope

Theoretically all activities that can reduce GHG emissions, energy consumption and pollution can be called low-carbon activities, including those in the field of low-carbon energy, low-carbon technology and low-carbon industries. A low-carbon energy system refers to the development of clean energy such as wind energy, solar energy, nuclear energy, geothermal energy and biomass energy to replace coal and petroleum, and eventually reduce CO_2 emission. Low-carbon technologies include IGCC and CCS. Low-carbon industries are built through the promotion of new energy automobiles, energy-saving buildings, industrial energy saving and recycling, resource recovery, environmental protection equipment, energy-saving material, etc.

Relevance

This indicator reveals how much money and effort a municipality is willing to put into low-carbon activities. In August of 2010, the NDRC launched low-carbon pilot projects in five provinces and eight cities in China. Low-carbon activities can help achieve environmental conservation as well as economic structure transformation and energy efficiency increase.

Instrument

$$\text{Ratio} = \frac{\text{Total fund spent on low-carbon activities}}{\text{Local government budget}}.$$

Data Sources

Municipal DRC, municipal financial budget department, Environmental Protection Bureau.

Data Sheet

Parameter	Unit	Quantity
Total fund spent on low-carbon activities	*Yuan*	
Local government budget	*Yuan*	

Key Responsible Entity for Action

Municipal financial department.

Action Checklist:

Item	Yes	No	N/A
Controlled Items			
Is financial support for low-carbon activities stated in the official governmental planning documents?			
Is there a earmarked budget for local low-carbon activities?			
Is the ratio of municipality budget on low-carbon activities increasing (double accounting avoided)?			
Is the scope of low-carbon activities funded by such budget expanding (double accounting avoided)?			
General Items			
Are there defined rules to regulate the management and usage of funds for low-carbon activities?			
Optional Advanced Items			
Has the local government set up channels to finance enterprises for adopting energy-saving technology?			
Is an evaluation system established to assess the effect of funds spent in promoting low-carbon development?			

3.1.2 *Implementation of Low-Carbon Planning and Policies*

(1) Incentives for Low-Carbon and Green Economy Promotion

Definition

This indicator is to evaluate the effort of a municipality to promote low-carbon and green economy. New energy, renewable energy technology and solar heating all need large amounts of capital input from the government. This indicator also reflects the production and consumption of decentralized energy. All incentives on the macro policy level for green economy promotion can be accounted.

Scope

Incentives for low-carbon and green economy promotion include financial incentives like subsidies, preferential loans and pricing and non-financial incentives like awards, encouragement of decentralized energy such as heat pumps, small biogas plants, solar heating, etc. Incentives for the application of enhanced efficiency standards in buildings, industries, land use planning and energy training system should also be accounted.

Relevance

Incentives for low-carbon and green economy promotion are very important as they can assist implementation of low-carbon policies. Enforcement of these policies is critical to achieve development goals.

Data Sources

Economy and Industry Committee, Bureau of Commerce, Low-Carbon Workgroup, DRC.

Key Responsible Entities for Action

Municipal DRC.

Action Checklist

Item	Yes	No	N/A
Controlled Items			
Are incentives for low-carbon and green economy promotion stated in local governmental planning documents?			
Are there financial incentives like subsidies, preferential loans and pricing available?			
Are there non-financial incentives like awards available?			
Has decentralized energy produced and consumed locally (like heat pumps, small biogas plants, solar heating) increased this year?			

(Continued)

(Continued)

Item	Yes	No	N/A
Is there supporting policy to help enterprises apply for national incentive programs on low-carbon activities?			
General items			
Are there any incentives for the application of enhanced efficiency standards in buildings, industries and land use planning?			
Optional advanced items			
Is there an energy training system established?			

(2) Number of Demonstrations in Low-Carbon Communities/ Schools/Hospitals/Shopping Malls/Supermarkets/Hotels, etc.

Definition

This indicator is to reflect the efficiency of low-carbon communities from the community service perspective, including the number of schools/hospitals/supermarkets, etc. At this phase an increase in the number is considered necessary.

Scope

The scope of this indicator covers low-carbon communities, schools, hospitals, shopping malls, supermarkets and hotels.

Relevance

This indicator is important to evaluate the efficiency of constituent parts in low-carbon communities. The government can put much effort into setting up low-carbon hospitals, low-carbon schools and low-carbon supermarkets, which helps make the whole community achieve low-carbon development.

Instruments

Local survey, municipal statistic yearbook.

Data Sources

Economy and Industry Committee, Bureau of Commerce, Low-Carbon Workgroup, DRC, local commercial office, local education bureau, local health bureau, etc.

Data Sheet

Parameter	Unit	Quantity
Number of demonstration hospitals in the city		
Number of demonstration schools in the city		
Number of demonstration supermarkets in the city		
Number of demonstration low-carbon communities in the city		
Number of demonstration shopping malls in the city		
Number of demonstration hotels in the city		

Key Responsible Entities for Action

Local commercial office, local education bureau, local health bureau.

Action Checklist

Item	Yes	No	N/A
Controlled Items			
Is setting up demonstrations of low-carbon communities/schools/hospitals/supermarkets/hotels stated in local governmental planning documents?			
Are there demonstrations of low-carbon communities/schools/hospitals/supermarkets/hotels in the city?			
Is there a standard or appraisal mechanism to select low-carbon communities/schools/hospitals/supermarkets/hotels?			
Is the number of low-carbon communities/schools/hospitals/supermarkets/hotels increasing?			
General Items			
Does the local government endeavor to promote low-carbon concepts and policies?			
Optional Advanced Items			
Do lighthouse projects have obvious effects among the public?			

(3) Green Procurement (e.g. Labels in Electrical Appliances)

Definition

Green procurement means government pre-emptive purchases of labeled products which promise less negative impact on the environment. In doing so, the government may encourage enterprises to improve the action on the environment, and then set an example of green consumption for the whole society.

Scope

. This indicator includes both government purchase and enterprise purchase. Both labeled products and non-labeled products with high energy efficiency performance are included. Products that are energy-saving, water-saving, low pollution, low toxicity, renewable and recyclable can all be classified as green products. The procurements using the public budget are all classified as government procurement. The list of green procurement refers to the national directory issued by the Ministry of Finance — Notice On Adjustments for List of Environment Labeled Products for Government Procurement (Document #: MOF[2008]50).

Relevance

This indicator is important to evaluate the government's effort to push enterprises to improve their low-carbon technologies. Consumer demand for low-carbon electronic products will put pressure on enterprises to adopt advanced technologies.

Instruments

Documents of municipal procurement policies.

Data Sources

Economy and Industry Committee, Bureau of Commerce, Low-Carbon Workgroup, DRC, Environmental Division, Ministry of Finance, local bureau of finance.

Data Sheet

Parameter	Unit	Quantity
Money spent by the municipality on buying products on the list in year 2010	*Yuan*	
Number of products sold out on the list in year 2010	Units	

Key Responsible Entities for Action

Low-Carbon Workgroup, Municipal DRC.

Action Checklist

Item	Yes	No	N/A
Controlled Items			
Is green procurement strategy stated in local governmental planning documents?			
Are there local incentives to encourage governmental departments and enterprises to buy products on the list, such as tax reduction, subsidy or privilege?			
Is there any regulation to guide the purchase of products in the list?			
General Items			
Is there clear regulation for government green procurement?			

(4) Information Disclosure/Accessibility for Low-Carbon Planning and Management Data

Definition

The indicator of information disclosure and accessibility for low-carbon planning and management documents can be obtained through regular government documents and other channels such as conferences, community bulletin boards, newspapers, etc.

Scope

This indicator includes all documents about low-carbon planning and management, not only at the central government level but also at the municipal level. The low-carbon strategy in the city's plan should cover all industries. The key words on a low-carbon economy in the Twelfth Five-Year Plan are "transformation of economic development patterns", "energy conservation and emission reduction", "setting up statistical accounting system of greenhouse gases emissions", "establishing market for carbon emission trading", "accelerate low-carbon technology development" and "broaden international cooperation".

Relevance

This indicator is important to connect low-carbon planning with the whole society, and increase everyone's low-carbon awareness. It will help provide clear future direction for enterprises and individuals.

Data Sources

Economy and Industry Committee, Bureau of Commerce, Low Carbon Office, DRC, municipal government work report.

Key Responsible Entities for Action

Low-Carbon Office and DRC.

Action Checklist

Item	Yes	No	N/A
Controlled Items			
Are there local information platforms (like websites or bulletin boards) for advertising low-carbon planning and management documents?			
Can information on low-carbon planning be published in a timely manner?			
Are coaching activities held regularly?			
Common Items			
Are there channels for the public to present their feedback on low-carbon planning and management?			
Optional Advanced Items			
Are there clear and transparent mechanisms for responding to public feedback?			

3.1.3 *Low-Carbon Management for Utilities*

(1) Energy Consumption in District Heating (tce/10000 m² · HDD)

Definition

This refers to the total energy consumption by district heating by heat suppliers during the reference period. Both steam heating and hot water heating are included.

Scope

Heat suppliers mainly refer to power plants and district boiler houses. The scope of energy consumption concerns the whole process including heat generation, transmission and heat loss. The types of energy consumed for heating include coal, oil, natural gas and electricity (e.g. pump water recycling). The scope of the heating period is defined by the local government. The overall scope of district heating refers to that for urban areas only.

Relevance

This indicator reflects the energy consumption and energy-saving potential of existing residential buildings in northern China. It is an important factor in municipal infrastructure and building energy efficiency.

Data Sources

Surveys need to be conducted for waste heat recovery and utilization in thermal power plants. The Housing and Urban-Rural Development Bureau, Statistics Bureau and heat suppliers will have some data. Basic statistics data on heating in cities can be found in the *China Urban and Rural Construction Statistical Yearbook* (compiled by Integrated Finance Division, Ministry of Housing and Urban-Rural Development).

Data Sheet

Parameter	Unit	Quantity
Total energy consumption	tce	
Total heating area	10,000 m²	
Heating period	day	
Energy consumption in district heating	tce/10000 m² · HDD	

Key Responsible Entity for Action

Municipality department responsible for heat supply.

Action Checklist

Item	Yes	No	N/A
Controlled Items			
Establish specific institution to be in charge of district heating			
Measures implemented to enforce energy regulations in district heating system			
Management scheme on district heating established			
General Items			
Targets are set for energy efficiency improvement			
Implementation plan to achieve the targets is developed			
Establish online monitoring system for heat supply			
Optional Advanced Items			
Set up detailed targets on energy consumption reduction for different phases including generation, transmission and heat loss.			

(2) Coverage Rate of Urban Population with Access to Gas (%)

Definition

Coverage rate of urban population with access to gas refers to the ratio of urban population with access to gas to the total urban population at the end of the reference period. The formula is:

$$\text{Coverage rate of urban population with access to gas} = \frac{\text{Urban population with access to gas}}{\text{Urban population}} \times 100\%$$

Scope

Gas includes coal gas, liquefied petroleum gas and natural gas. Volume of gas supplied refers to the total volume of gas provided to users by gas-producing enterprises in a year, including the volume sold and the volume lost.

Relevance

It is one of the most important factors in the municipal infrastructure and reflects the municipal infrastructure development level.

Data Sources

Housing and Urban-Rural Development Bureau, Statistics Bureau, gas company.

Basic statistics data on gas supplied can be found from the *China Urban and Rural Construction Statistical Yearbook* (compiled by Integrated Finance Division, Ministry of Housing and Urban-Rural Development).

Data Sheet

Parameter	Unit	Quantity
Urban population with access to gas	Person	
Total urban population	Person	

Action Checklist

Item	Yes	No	N/A
Controlled Items			
Establish specific institution in charge of district gas supply			
General Items			
Targets are set for increasing gas coverage			
Implementation plan to achieve the targets is established			
Online monitoring system for gas supply established			
Optional Advanced Items			
Set up detailed targets on energy consumption reduction for different phases including gas production, transmission and pipe loss.			

(3) Share of Local Renewable Energy Production in Total Energy Consumption (%)

Definition

Share of local renewable energy production in total energy consumption.

Scope

According to China's Renewable Energy Law, renewable energy refers to non-fossil energy such as wind energy, solar energy, hydro energy, biomass energy, geothermal energy and oceanic energy. The government makes the development and utilization of renewable energy a priority for energy development and promotes the development of the renewable energy market by establishing the total volume for renewable energy development and taking corresponding measures. The government encourages economic entities of all ownerships to participate in the development and utilization of renewable energy and protect the legal rights and interests of the developers and users of renewable energy.

$$\text{Share of local renewable energy production in total energy consumption} = \frac{\text{Local renewable energy production}}{\text{Total energy consumption}}$$

Relevance

Carbon emissions are largely caused by fossil energy consumption. The emission factor of oil is lower than that of coal and higher than that of natural gas. Green plants are carbon-neutral, while renewable energy is considered clean energy with zero carbon emission. The energy mix can be evaluated by two indicators: energy carbon intensity (i.e. carbon emission per unit energy consumed, which reflects energy consumption structure) and proportion of zero-carbon energy (including renewable energy and nuclear energy) in total primary energy consumption. The development of renewable energy is subject to resource endowment, funding and technologies. It is crucial to a low-carbon economy and low-carbon development. Considering though that some renewable energy is used for grid power generation, not all for local consumption, it can still reflect the contribution to national renewable energy development.

Data Sources

Municipal DRC, Statistics Bureau.

Data Sheet

Parameter	Unit	Quantity
Total energy consumption	tce	
Local renewable energy production	tce	

Key Responsible Entity for Action

Municipal DRC, Statistics Bureau and Economic and Trade Bureau.

Action Checklist

Item	Yes	No	N/A
Controlled Items			
Establish specific institution to be in charge of local renewable energy production			
Measures established to enforce local renewable energy development			
General Items			
Targets are set for renewable energy production			
Implementation plan is set to achieve the targets defined			
Optional Advanced Items			
Set up detailed targets on local renewable energy production for different fields including renewable energy used locally, and that transmitted to other places			
Develop targets and timetable for renewable energy development			

(4) Energy Consumption for Water Supply (tce/m³)

Definition

Energy consumption of water supply in urban areas refers to the total energy consumption associated with water supplied to urban users by water utilities during the reference period.

Scope

Production capacity of water supply refers to the designed overall production capacity of water facilities, covering the four segments of water collection, purification, conveyance and outflow through trunk pipelines. Increased capacity through transformation and innovation projects is included as well. The capacity is determined mainly by the weakest section of the above-mentioned four.

Residential use water consumption refers to water consumption of households for daily life and water consumption of public service facilities. The latter refers to water consumption for urban public services, including government agencies, public institutions, military barracks, public facilities, wholesale and retail outlets, restaurants, hotels and other entities providing public services.

Household water consumption refers to consumption of water for daily life of all households within the boundary of cities, including households of urban residents and farmers, and public water supply stations.

Relevance

It is one of the most important factors in the municipal infrastructure.

Data Sources

Data on water supply can be from the Water Bureau, Statistics Bureau and water supply entities. Data on energy consumption associated with water supply can be from water supply entities and property management agencies.

Action Checklist

Item	Yes	No	N/A
Controlled Items			
Targets set for energy consumption associated with water supply			
Implementation plan established to achieve the targets defined			
Supervision on the operation of online monitoring system for energy consumption associated with water supply			
General Items			
Establish specific institution to be in charge of energy consumption associated with water supply			
Measures established to enforce policies on energy consumption associated with water supply			
Optional Advanced Items			
Set up detailed targets on energy consumption associated with water supply for different process			

(5) Energy Consumption for Sewage Water Treatment (tce/m^3)

Definition

Energy consumption associated with industrial waste water and municipal waste water treatment refers to the total energy consumption associated with water treatment by industrial enterprises and waste water treatment plants during the reference period.

Scope

Industrial waste water includes waste water from the production process, cooling water, groundwater from mining wells and sewage from households. Municipal waste water refers to waste water produced by urban households. This indicator covers energy consumed for the treatment of industrial waste water and municipal waste water, excluding the energy consumption of the industrial production process itself.

Relevance

It reflects the energy efficiency and low-carbon development level of the municipal infrastructure and industrial production infrastructure.

Data Source

Data on energy consumption associated with industrial waste water can be from the Municipal Environmental Protection Bureau, Statistics Bureau and industrial enterprises. Data on energy consumption associated with municipal waste water can be from the Municipal Public Utility Bureau, Statistics Bureau and waste water treatment plant.

Data Sheet

Parameter	Unit	Quantity
Total energy consumption	tce	
Total quantity of treated waste water	m^3	

Action Checklist

Item	Yes	No	N/A
Controlled Items			
Targets set for energy consumption associated with waste water treatment			
Implementation plan established to achieve the targets defined			
General Items			
Establish specific institution to be in charge of energy consumption associated with waste water treatment			
Measures established to enforce policies on energy consumption associated with waste water treatment			
Optional Advanced Items			
Set up detailed targets on energy consumption associated with waste water treatment for different process			

(6) Water Conservation Measures in Municipal Water Management

Definition

Water conservation refers to reduction in water consumption and increases in waste water recycling for different usages such as cleaning and manufacturing.

Scope

The goals of water conservation efforts include the following: (1) Sustainability. To ensure water availability for future generations, the withdrawal of fresh water from an ecosystem should not exceed its natural replacement rate. (2) Energy conservation and de-carbonization. Water pumping, delivery and waste water treatment facilities consume a significant amount of energy. In some regions of the world over 15% of total electricity consumption is devoted to water management.

Water conservation programs are typically initiated at the local level, by either municipal water utilities or regional governments. Common strategies include public outreach campaigns, tiered water rates (charging progressively higher prices as water use increases), or restrictions on outdoor water use such as lawn watering and car washing. Cities in dry climates often require or encourage the installation of natural landscaping in new homes to reduce outdoor water usage.

Relevance

Water conservation can be defined as:

(1) Any beneficial reduction in water loss, use or waste as well as the preservation of water quality.
(2) A reduction in water use accomplished by implementation of water conservation or water efficiency measures.
(3) Improved water management practices that reduce or enhance the beneficial use of water. A water conservation measure is an action, behavior change, device, technology, or improved design or process implemented to reduce water loss, waste or use. Water

efficiency is a tool of water conservation. That results in more efficient water use and thus reduces water demand. The value and cost-effectiveness of a water efficiency measure must be evaluated in relation to its effects on the use and cost of other natural resources (e.g. energy or chemicals).

Data Sources

Municipal Public Utility Bureau, Water Bureau, Statistics Bureau.

Action Checklist

Item	Yes	No	N/A
Controlled Items			
Establish specific institution to be in charge of energy consumption associated with water supply			
Measures established to enforce policies on energy consumption associated with water supply			
General Items			
Targets are set for water conservation			
Implementation plan established to achieve the targets defined			
Optional Advanced Items			
Build pilot projects in water conservation measures, and encourage industrial standards in water conservation measures			

(7) Treatment Rate of Municipal Solid Waste (%)

Definition

Municipal solid waste (MSW) refers to solid waste generated in everyday life activities and in service-providing activities for urban life, including: household waste, waste generated in market activities, cleaning of streets, public sites, offices, schools, factories, mines, etc.

The treatment rate of municipal solid waste refers to municipal solid waste treated over that produced. In practical statistics, as it is difficult to estimate, the volume of municipal solid waste produced is replaced with the collected volume. It is calculated as:

$$
\text{Treatment rate of municipal solid waste}
$$
$$
= \frac{\text{Municipal solid waste treated}}{\text{Municipal solid waste produced}} \times 100\%
$$

Scope

Municipal solid waste predominantly includes food waste, yard waste, containers and product packaging, and other miscellaneous inorganic waste from residential, commercial, institutional and industrial sources. Examples of inorganic waste are appliances, newspapers, clothing, food scraps, boxes, disposable tableware, office and class-room paper, furniture, wooden pallets, rubber tires and cafeteria waste. Municipal solid waste does not include industrial waste, agricultural waste and sewage sludge. The collection is performed by the municipality within a given area. They are in either solid or semisolid form. The term "residual waste" relates to waste left from household sources containing materials that have not been separated out or sent for reprocessing. Following are the different types of waste:

- Biodegradable waste: food and kitchen waste, green waste, paper (can also be recycled).
- Recyclable material: paper, glass, bottles, cans, metals, certain plastics, etc.
- Inert waste: construction and demolition waste, dirt, rocks, debris.
- Domestic hazardous waste (also called "household hazardous waste") and toxic waste: medication, e-waste, paints, chemicals, light bulbs, fluorescent tubes, spray cans, fertilizer and pesticide containers, batteries, shoe polish cream.

Relevance

The municipal solid waste industry has four components: recycling, composting, land filling and waste-to-energy via incineration. The primary steps are generation, collection, sorting and separation, transfer and disposal.

Waste generation encompasses activities in which materials are identified as no longer being of value and are either thrown out or gathered together for disposal.

The functional element of collection includes not only the gathering of solid waste and recyclable materials, but also the transport of these materials, after collection, to the location where the collection vehicle is emptied. This location may be materials processing facility, a transfer station or a landfill disposal site.

Waste disposal and classification, storage and processing at the source: Waste disposal and classification involve activities associated with waste management until the waste is placed in storage containers for collection. Handling also encompasses the movement of loaded containers to the point of collection. Separating different types of waste components is an important step in the handling and storage of solid waste at the source.

Classification and processing and transformation of solid waste: The types of means and facilities that are now used for the recovery of waste materials that have been classified at the source include curb side collection, drop off and buy back centers. The separation and processing of waste that has been separated at the source and the separation of commingled waste usually occur at a materials recovery facility, transfer stations, combustion facilities and disposal sites.

Transfer and transport: This element involves two main steps. First, the waste is transferred from a smaller collection vehicle to larger transport equipment. The waste is then transported, usually over long distances, to a processing or disposal site.

Disposal: Today the disposal of waste by land filling or land spreading is the ultimate fate of all solid waste, whether it is residential waste collected and transported directly to a landfill site, residual materials from materials recovery facilities (MRFs), residue from the combustion of solid waste, compost or other substances from various solid waste processing facilities. A modern sanitary landfill is not a dump; it is an engineered facility used for disposing of solid waste on land without creating nuisances or hazards to public health or safety, such as the breeding of insects and the contamination of ground water.

Energy generation: The urban solid waste can be used to generate energy. Several technologies have been developed that make the processing of municipal solid waste for energy generation cleaner and more economical than ever before, including landfill gas capture, combustion, gasification and plasma arc gasification. While older waste incineration plants emitted high levels of pollutants, recent regulatory changes and new technologies have significantly reduced this concern.

Data Sources

Municipal Public Utility Bureau, Statistics Bureau.

Action Checklist

Item	Yes	No	N/A
Controlled Items			
Establish specific institution to be in charge of energy consumption associated with waste treatment			
Measures established to enforce policies on energy consumption associated with waste treatment			
General Items			
Targets are set for waste treatment			
Implementation plan established to achieve the targets defined			
Optional Advanced Items			
Build demonstration pilot in waste treatment, and encourage industrial standards in waste treatment			

(8) Municipal Solid Waste Sorting and Reduction

Definition

Measures of municipal solid waste sorting and reduction are the collection, transport, processing, recycling or disposal, and monitoring of waste materials. They usually relate to materials produced by human activity, and are generally undertaken to reduce their effect on health, the environment or aesthetics. Measures of municipal solid waste sorting and reduction are also carried out to recover resources from them. Measures of municipal solid waste sorting and reduction can involve solid, liquid, gaseous or radioactive substances, with different methods and fields of expertise for each.

Scope

Measures of municipal solid waste sorting and reduction attempt to offer the most benign options for waste management. For mixed municipal solid waste a number of broad studies have indicated waste administration, then source separation and collection followed by reuse and recycling of the non-organic fraction and energy and compost/fertilizer production of the organic waste fraction via anaerobic digestion to be the favored path. Non-metallic waste resources are not destroyed as with incineration, and can be reused/recycled in a future resource-depleted society.

An important method of municipal solid waste sorting and reduction is the prevention of waste material being created, also known as waste reduction. Methods of avoidance include reuse of second-hand products, repairing broken items instead of buying new, designing products to be refillable or reusable (such as cotton instead of plastic shopping bags), encouraging consumers to avoid using disposable products (such as disposable cutlery), removing any food/liquid remains from cans, and packaging and designing products that use less material to achieve the same purpose.

Relevance

There are a number of concepts about the measurement of municipal solid waste sorting and reduction which vary in their usage between countries or regions.

The waste hierarchy refers to the "3Rs" — reduce, reuse and recycle — which classify waste management strategies according to their desirability in terms of waste minimization. The waste hierarchy remains the cornerstone of most waste minimization strategies. The aim of the waste hierarchy is to extract the maximum practical benefits from products and to generate the minimum amount of waste.

Extended producer responsibility (EPR) is a strategy designed to promote the integration of all costs associated with products throughout their life cycle (including end-of-life disposal costs) into the market price of the product. EPR is meant to impose accountability over the entire life cycle of products and packaging introduced to the market. This means that firms which manufacture, import and/or sell products are required to be responsible for the products after their useful life as well as during manufacture.

Polluter pays principle: This is a principle where the polluting party pays for the negative impacts made by polluters. With respect to waste management, this generally refers to the requirement for a waste generator to pay for appropriate disposal of the waste.

Data Sources

Municipal Public Utility Bureau, Statistics Bureau.

Action Checklist

Item	Yes	No	N/A
Controlled Items			
Establish specific institution to be in charge of municipal solid waste sorting and reduction			
Residents have high level of awareness about municipal solid waste sorting and reduction			
Proper infrastructures are in place for municipal solid waste sorting and reduction			
General Items			
Supervision and management targets are set for municipal solid waste sorting and reduction measures			
Implementation plan, such as incentives or penalties, to achieve the targets is defined			
Demonstration pilots for municipal solid waste sorting and reduction measures			
Optional Advanced Items			
Municipal solid waste sorting and reduction measures are effective			
Handbooks and guidelines set out municipal solid waste sorting and reduction measures are available			

3.2 Low-Carbon Economy

3.2.1 *Low-Carbon Industries*

(1) Non-CO_2 GHG Emission in Industries and Mitigation Measures

Definition

Non-CO_2 GHG emissions and their mitigation measures in industrial process refer to non-CO_2 GHGs emissions in municipal industrial process measured by emission inventory tools and corresponding measures planned and implemented to mitigate these emissions by municipality and industrial enterprises.

Scope

GHG emissions by a municipality refers to the emission of six main types of greenhouse gases outlined by the Kyoto Protocol, i.e. CO_2, CH_4, N_2O, HFCs, PFCs and SF_6. Since the main indicators measure CO_2 and CH_4, this indicator focuses on non-CO_2 greenhouse gases, e.g. N_2O, HFCs, PFCs and SF_6 in industrial process, which have much larger global warming potential (GWP) compared with CO_2. This indicator does not include residential carbon emissions or carbon emission embedded in industrial materials.

The inventory tool typically is a spreadsheet or database to help the city to calculate GHG emissions of different sources. The UNFCCC defines the inventory tool as "a type of emission inventory developed for a variety of reasons". The EPA defines it as an interactive spreadsheet model designed to help GHG emissions inventories. It is a top-down approach to calculate estimates of GHG emissions. It looks at direct emissions only, provides a streamlined way to update an existing inventory or complete a new inventory. It allows users to create a simple forecast of emissions. Inventory tools for municipalities can be developed by local authorities with professional consultancies when necessary.

Mitigation measures include both existing and new technologies and methodologies which can mitigate or reduce GHGs emissions directly or help facilitate emission mitigation in municipal industries, e.g. adopting low-carbon technologies, phasing out carbon-intensive technologies, optimizing industrial process flow, using renewable energy, restructuring and upgrading industries, and so on.

Relevance

It is to reflect the direct non-CO_2 GHG emissions mitigation in the industrial sector.

According to the IPCC TAR (2001) and IPCC AR4 (2007), the six main types of greenhouse gases listed by the Kyoto Protocol have different GWPs, which compares the amount of heat trapped by a certain mass of the gas in question to the amount of heat trapped by a similar mass of carbon dioxide. Since CO_2 and CH_4 have been measured by LCCC main indicators, this supporting indicator supplements the remaining four GHGs, which have much bigger GWPs (see Table 3.1). By checking the non-CO_2 GHGs in the industrial process, this indicator helps to increase the municipalities' awareness of their GHG emission status and prompts them to adopt appropriate mitigation measures steadily.

Table 3.1   GWPs of Different GHGs.

GHG	20 years	100 years	500 years
CO_2	1	1	1
CH_4	62	27	7
N_2O	275	310	256
PFCs	9,400	11,700	10,000
HFCs	3,900	5,700	8,900
SF_6	15,100	22,200	32,400

Data Sources

Municipal DRC, Statistical Bureau.

Instrument

A carbon emission inventory is needed, which subdivides municipal industries according to different GHGs emissions. The inventory should be updated annually based on regular measurement, monitoring and calculation. A systematic guidance on industrial mitigation measures should be provided.

Key Responsible Entities for Action

Municipal DRC, Economic and Industrial Committee, main industrial enterprises.

Action Checklist

Item	Yes	No	N/A
Controlled Items			
Establish necessary inventory tools for GHGs, including six types of main gases, in industrial sector			
Non-CO_2 GHG emissions in industrial sector regularly monitored			
General Items			
Appropriate measures planned and implemented to mitigate non-CO_2 GHGs emissions in industrial sector			
Targets are set for mitigating non-CO_2 GHGs in industrial sector			
Implementation plan to achieve the targets is defined			
Optional Advanced Items			
Good practice or pilot projects in mitigating non-CO_2 GHG in industrial sector implemented			

(2) Implementation of Investment Policies for Energy-Intensive and Highly Polluting Industries

Definition

This indicator reflects that when carrying out investment policies, how strict the municipality is on the entry of capital/companies which have been categorized as energy-intensive and highly polluting industries.

Scope

Energy-intensive and highly polluting industries refer to those listed in the national directory of investment policies for energy-intensive and highly polluting industries by the NDRC, Notice of the State Council on Further Strengthening the Elimination of Backward

Production Capacities (GuoFa [2005]40) and Notice of the State Council on Accelerating Structural Adjustment of Overcapacity Industries (GuoFa [2006]11). The municipality can have stricter policies according to the local industrial structure.

Relevance

This indicator is to reflect the efforts on restructuring and upgrading the municipal industry towards a low-carbon economy.

To identify and reject energy-intensive and highly polluting industries in local investment is an important approach to control carbon emissions in new investment and restructure a city's low-carbon economy. This indicator could reflect the efficiency of local carbon management on implementing the NDRC's structural adjustment policies on elimination of backward production capacities.

Data Sources

DRC, Environmental Protection Bureau.

Instruments

Documents of municipal investment policies, field surveys.

Key Responsible Entities for Action

Municipal DRC, Economic and Industrial Committee.

Action Checklist

Item	Yes	No	N/A
Controlled Items			
Set a strategic plan to implement national targets to control the rapid growth of energy-intensive and highly polluting industries and to enforce structural adjustment of overcapacity industries			
Newly constructed and expanded energy-intensive and highly polluting projects are checked, verified and recorded according to the entrance criteria			

(*Continued*)

(*Continued*)

Item	Yes	No	N/A
Measures and policies to support and encourage technological improvement on energy-saving and carbon emission mitigation			
Measures on power demand side management implemented in energy-intensive and highly polluting industries, including establishing special funds for power demand side management, applying TOU tariff, etc. A proper power program should be developed (a proper power program means taking legal, administrative, economic, technical and other means to enhance power management, changing users' behavior to shift load, averting holidays, reducing loads, negative control, etc., to properly manage power consumption and avoid imposed blackouts which bring adverse impacts on communities and businesses). A proper power program should be initiated by the government and involve utilities and users. It is necessary to avoid imposed blackouts and reduce interruption for production of key enterprises and life in communities			
The national pollutant emission standards are implemented in energy-intensive and highly polluting enterprises and the noncompliant enterprises have taken corrective measures			
General Items			
Whether newly invested projects in energy-intensive and highly polluting industries implement energy efficiency assessment, environmental impact assessment, including project approval, case filing and whether appropriate project accountability is established			
Optional Advanced Items			
The municipality sets and implements stricter policies than national standards on the entry of energy-intensive and highly polluting industries			

(3) Eliminating Obsolete Capacity

Definition

Eliminating obsolete capacity refers to the efforts to gradually close down facilities with obsolete technologies in industries in order to restructure and upgrade the industrial structure towards a more low-carbon economy.

Scope

Obsolete technologies in industries are usually those with high pollution and high energy consumption. For details, refer to the Notice of the State Council on Accelerating Structural Adjustment of Overcapacity Industries (GuoFa [2006]11) and the Notice of the State Council on Further Strengthening the Elimination of Backward Production Capacities (GuoFa [2005]40). The municipality can have stricter policies according to the local industrial structure.

Relevance

This indicator is to reflect the efforts on restructuring and upgrading the municipal industry towards a low-carbon economy.

Data Sources

DRC, Environmental Protection Bureau.

Instruments

Documents of municipal investment policies, field surveys.

Key Responsible Entities for Action

Municipal DRC, Economic and Industrial Committee.

Action Checklist

Item	Yes	No	N/A
Controlled Items			
Establish special funds in local finance to support eliminating obsolete capacity			
Concrete measures taken by the municipality to support superior enterprises and phase out inferior ones through project approval			
General Items			
Set a plan (integrated into the official planning system) for phasing out obsolete capacity during the 12th Five-Year Plan period			
Optional Advanced Items			
The municipality sets and implements stricter policies on eliminating obsolete capacity than national standards			

(4) Energy Efficiency Compliance Rate for Industrial Enterprises Above Designated Size (%)

Definition

The energy efficiency compliance rate for industrial enterprises above the designated size refers to the percentage of industrial enterprises which have passed the evaluation of energy efficiency investigation conducted by designated institutions.

Scope

Industrial enterprises above the designated size refer to enterprises with a turnover from the main business of over RMB 5 million (2007–2010 standard), then RMB 20 million (standards since January of 2011) as defined by the National Statistic Bureau.

A regular monitoring system should be established for industrial enterprises above the designated size, which can adopt the provincial monitoring system or be developed by municipality.

Relevance

A regular investigation and evaluation mechanism should be enforced to ensure the large-scale industrial enterprises comply with the energy efficiency standards required by provinces and municipalities.

Data Sources

Economic and Industrial Committee, DRC.

Instrument

$$\text{Compliance rate} = \frac{\begin{array}{c}\text{Number of industrial enterprises above} \\ \text{designated size complying with energy} \\ \text{efficiency monitoring systemit}\end{array}}{\begin{array}{c}\text{Total industrial enterprises} \\ \text{above designated size}\end{array}}.$$

Data Sheet

Parameter	Unit	Quantity
Industrial enterprises above designated size complying with energy efficiency requirements	Company	
Total industrial enterprises above designated size	Company	

Key Responsible Entities for Action

Municipal DRC, Economic and Industrial Committee.

Action Checklist

Item	Yes	No	N/A
Controlled Items			
Establish and implement energy efficiency monitoring and investigation system for industrial enterprises in municipality			
General Items			
Results of energy efficiency monitoring and investigation are publicized regularly and in a timely manner			

(Continued)

(Continued)

Item	Yes	No	N/A
Enterprises which fail to pass energy efficiency monitoring requirements receive warning and guidance for improvement			
Optional Advanced Items			
An integrated energy efficiency monitoring system established and implemented, which pioneering peer municipalities or provinces			

(5) EMS Application Rate Among Priority Enterprises for Energy Efficiency (%)

Definition

This refers to the percentage of industrial enterprises which have established an energy management system (EMS) among all the major energy consumption enterprises in the municipality.

Scope

A national priority enterprise for energy efficiency is defined as that with a total energy consumption of over 10,000 tce/year (Energy Conservation Law of PRC), while for a provincial priority enterprise for energy efficiency the standard is 5,000–10,000 tce/year. Municipalities can enlarge the monitoring scope and lower the threshold for a priority enterprise for energy efficiency accordingly, e.g. from 2,000 to 5,000 tce/year.

The Energy Management System Requirements [GB/T23331-2009] came into effect on November 1, 2009. Through integrating energy conservation planning, energy conservation technology innovation, statistics, computation and trainings, etc., an energy management system helps researchers form an integral system and achieve energy-saving objectives. In addition, the standard introduces the idea of benchmarking and sets specific requirements for designing, manufacturing, external information collection, statistics and computation, rewards and punishment.

Relevance

This encourages major energy-consuming industrial enterprises to build up EMSes and pursue scientific energy management, which can help increase energy efficiency in industries.

Data Sources

Economic and Industrial Committee, DRC.

Data Sheet

Parameter	Unit	Quantity
Number of priority enterprises for energy efficiency industrial enterprises which have established an EMS		
Total number of priority enterprises for energy efficiency		

Key Responsible Entities for Action

Economic and Industrial Committee, DRC.

Action Checklist

Item	Yes	No	N/A
Controlled Items			
Institution exists to take charge of the promotion of an EMS among priority enterprise for energy efficiency			
Measures available to help establish an EMS in industrial enterprises			
General Items			
Technological and informational support, e.g. workshops, training courses, professional consultancies, are provided for enterprises on the EMS's establishment and operation			
Optional Advanced Items			
Assessment conducted on energy conservation effects for enterprises in implementing an EMS			
Above-mentioned effects disclosed or published to interested parties			
The rate reaches 90% and above			

(6) Number of Priority Enterprises for Energy Efficiency That Have Signed Voluntary Agreements for Emission Reduction

Definition

This indicator evaluates the number of priority enterprises for energy efficiency that do not have assigned obligations to reduce energy consumption but voluntarily signed agreements to do so.

Scope

A national priority enterprise for energy efficiency is defined as that with a total energy consumption of over 10,000 tce/year (Energy Conservation Law of PRC), while for a provincial priority enterprise for energy efficiency the standard is 5,000–10,000 tce/year. Municipalities can enlarge the monitoring scope and lower the threshold for a priority enterprise for energy efficiency accordingly, e.g. from 2,000 to 5,000 TCE/year.

Municipalities can design a template for a Voluntary Energy Conservation Agreement based on the sample provided by the central government (refer to General Technical Rules for Voluntary Agreement of Energy Conservation (2010), initiated by the NDRC and drafted by China's National Institute of Standardization).

Relevance

Besides the industrial enterprises which have binding mitigation obligations required by national, provincial or municipal authorities, there are large quantities of enterprises which do not have binding force or energy-saving obligation due to the comparatively small scale. Signing voluntary agreements thus is a recommendable way for local governmental authorities to guide the energy-saving/carbon emission mitigation behaviors in these enterprises.

Data Sources

Economic and Industrial Committee, DRC.

Key Responsible Entity for Action

Economic and Industrial Committee.

Action Checklist

Item	Yes	No	N/A
Controlled Items			
A voluntary agreement system in operation for energy-saving/carbon emission mitigation for industrial enterprises in the municipality			
General Items			
Activities organized to promote the voluntary energy-saving/carbon emission mitigation agreement system			
Policies developed to encourage industrial enterprises to sign a voluntary agreement			
Mechanism exists to evaluate actual effects			

(7) Ratio of Renewable Energy/Energy-Efficient Appliances' Product Value in Total Industrial Outputs

Definition

This indicator evaluates the ratio of renewable energy and energy-efficient appliances' product value in total industrial outputs.

Scope

Renewable energy appliances refer to the production of industrial commodities which are used as infrastructure, equipment, appliances, tools, etc., of renewable energy application, or help to apply renewable energies.

According to the Renewable Energy Law, renewable energy refers to non-fossil energy such as wind energy, solar energy, hydro energy, biomass energy, geothermal energy and oceanic energy.

Energy-efficient appliance certification refers to the certification activities conducted according to national standards and through internationally recognized procedures. The certification is supervised and guided by the General Administration of Quality Supervision, Inspection and Quarantine of China.

Energy-efficient appliances refer to the appliances which can reach quality and safety standards. Compared with similar products or products that perform the same function, the efficiency or energy-consuming indicators of these products can reach or is close to international and national advanced standards. According to the Law of the People's Republic of China on Conserving Energy and Regulations on Energy Efficient Appliances Certification Management, a product can only be labeled "energy-efficient" if it has been certified by relevant national authorities. Currently, there are national energy-efficient certification organizations and some provincial energy-efficient certification organizations. For more details please refer to the official website of the China Standard Certification Centre (CSC) (http://www.cecp.org.cn).

Relevance

This indicator is to reflect the production ability of renewable energy appliances and energy-efficient appliances manufacture. It is to encourage the enterprises in municipalities to pay attention to energy saving and carbon emission reduction from the production party by producing more renewable energy-related products and energy-efficient appliances.

Data Sources

The NDRC, China Standard Certification Centre (CSC, formerly CECP).

Instrument

Ratio

$$= \frac{\text{Renewable energy and energy-efficient appliances' product value}}{\text{Total industrial output}}.$$

Key Responsible Entity for Action

Economic and Industrial Committee.

Action Checklist

Item	Yes	No	N/A
Controlled Items			
Catalogue in place for energy-efficient and renewable energy products			
Statistics on renewable-energy-related products and energy-efficient appliance production in municipal industries are regularly carried out and updated			
General Items			
Strategic plan set for promoting renewable-energy-related products and energy-efficient appliance production			
Policies to support and encourage the production of renewable-energy-related products and energy-efficient appliances			
Optional Advanced Items			
The clear target of the ratio is set to foster renewable-energy-related products to a pivotal industry			

(8) The Application of Renewable Energy in the Industrial Sector

Definition

The application of renewable energy in the industrial sector refers to the share of renewable energy in the grid from where industrial enterprises purchase electricity and the renewable energy produced and consumed in industrial enterprises locally while this part of renewable energy does not feed into the national grid.

Scope

Since the majority of the municipal electricity can only be purchased from the national/provincial grid, the share of renewable energy in the grid on the municipal level is usually seen as a fixed mix of thermal power, hydro, renewable energy, nuclear power and so on.

According to China's Renewable Energy Law, renewable energy refers to non-fossil energy including wind energy, solar energy, hydro energy, biomass energy, geothermal energy, oceanic energy, etc. Renewable energies produced and consumed in industrial enterprises locally vary greatly according to resource endowment among municipalities. Here this indicator tends to reflect all kinds of application of renewable energy in the municipality, taking into consideration the various forms of energy listed above.

Relevance

Application of renewable energy is a key approach to realize low-carbon development globally. It is one of the main targets set in China's medium- and long-term energy plan as well, which calls for increasing the share of renewable energy from the current 7% to 10% by 2010 and further to 15% by 2020. On the municipal level, different forms of innovation and expansion of renewable energy should be encouraged including not only power generation, but also solar heating, geothermal heating bumps, biogas for cooking and lighting, etc.

Data Source

Economic and Industrial Committee, Statistic Bureau, DRC.

Key Responsible Entity for Action

Economic and Industrial Committee.

Action Checklist

Item	Yes	No	N/A
Controlled Items			
A specific target for renewable energy application set for the industrial sector			
General Items			
Policies to encourage and support renewable energy application by industrial enterprises			
Optional Advanced Items			
Good practice in renewable energy application in industries pioneering in peer municipalities or provinces			

(9) Recycling and Utilization Rate of Industrial Solid Waste (%)

Definition

The rate of utilization of industrial solid waste refers to the percentage of industrial solid waste utilized over industrial solid waste produced (including stocks of the previous years). According to the *Statistical Yearbook 2010*, industrial solid waste utilized refers to the volume of solid waste from which useful materials can be extracted or which can be converted into usable resources, energy or other materials by means of reclamation, processing, recycling and exchange (including utilizing in the year the stocks of industrial solid waste of the previous year).

Scope

Examples of utilizations of industrial solid waste include fertilizers, building materials and road materials. Industrial solid waste produced refers to the total volume of solid, semi-solid and high concentration liquid residues produced by industrial enterprises in the production process in a given period of time, including hazardous waste, slag, coal ash, gangue, tailings, radioactive residues and other waste, but excluding stones stripped or dug out in mining — gangue and acid or alkaline stones not included (a stone is acid or alkaline according to the pH value of the water being below 4 or above 10.5 when the stone is in or soaked by water).

Hazardous waste is waste that poses substantial or potential threats to public health or the environment as listed in the national inventory of hazardous waste or according to characteristics identified by national standards as hazardous waste, e.g. ignitable, explosive, reactive, corrosive, toxic or carrying infectious diseases.

Relevance

Industrial waste takes the form of three kinds of wastes: waste gas, waste water and solid waste. The treatment and utilization of solid waste is a crucial factor for environmental protection in industrial production and can represent the comprehensive utilization technologies and capabilities in municipal industries.

Data Source

Environmental Protection Bureau, *Statistical Yearbook.*

Instrument

Rate of utilization of industrial solid wastes

$$= \frac{\text{Volume of industrial solid wastes utilized}}{\text{Industrial solid wastes produced} + \text{stock of previous year}} \times 100\%$$

Key Responsible Entities for Action

Economic and Industrial Committee (Environmental Protection Bureau is in charge of hazardous waste and medical waste).

Action Checklist

Item	Yes	No	N/A
Controlled Items			
The treatment and utilization of industrial solid waste is regulated and regularly supervised by the government			
General Items			
Regularly compile and publish Industrial Solid Waste Utilization Report			
Optional Advanced Items			
Good practices and advanced technologies recommended and encouraged to apply in industries by the government			

3.2.2 *Low-Carbon Service*

(1) Number of Trained Energy Managers in Hospitals, Schools, Large Shopping Malls, Hotels, Airports, etc.

Definition

This indicator is to require the public service institutions to be equipped with energy management staff with professional training experience. It can be reflected through the number of energy managers trained in the EMS by relevant institutions, governmental departments or in internal training programs.

Scope

This indicator involves all public service institutions in municipalities, including hospitals, schools, large shopping malls, hotels, airports, railway stations and so on.

The Energy Management System Requirements [GB/T23331-2009] came into effect on November 1, 2009. Through integrating energy conservation planning, energy conservation technology innovation, statistics, computation and trainings, etc., the EMS helps managers form an integral system and achieve energy-saving objectives. In addition, the standard introduces the ideology of benchmarking and sets specific requirements for design, manufacturing, external information collection, statistics and computation, rewards and punishment.

Relevance

In the service sector, human resources and expertise on energy management are a crucial factor in increasing energy saving and carbon emission reductions. This indicator is to promote the efforts in public service institutions in adopting and training professional energy management staff.

Instrument

Local surveys.

Data Sources

Economic and Industrial Committee, Bureau of Commerce, Low-Carbon Workgroup, DRC.

Key Responsible Entities for Action

Bureau of Commerce, Economic and Industrial Committee.

Action Checklist

Item	Yes	No	N/A
Controlled Items			
Full-time energy management staff/specialist in service institutions with proper TOR			
Regular EMS training courses are carried on in service institutions and enterprises			
General Items			
Regular experience-exchanging activities, such as workshops, seminars in the public service sector			
Institutions assigned to be responsible for and monitoring the process of the EMS training course			
Optional Advanced Items			
The training records of energy managers are documented			

(2) Ratio of Major Service Industries That Signed Voluntary Agreements for Emission Reduction

Definition

This indicator is to reflect the percentage of entities in the service sector with no binding mitigation obligation but sign agreements with government authorities to reduce carbon emissions or energy consumption voluntarily.

Scope

The scope of major service industries in a municipality can be defined according to the local service sector condition. Three to five main service industries need to be covered. This indicator intends to involve large-scale service enterprises above a certain level with thresholds defined locally, but small-to-medium enterprises are also encouraged to join in.

Municipalities can design a template for the Voluntary Energy Conservation Agreement based on the sample provided by the central government (refer to General Technical Rules for Voluntary Agreement of Energy Conservation (2010), initiated by the NDRC and drafted by China's National Institute of Standardization).

Relevance

In the service sector, the enterprises usually do not have binding emission mitigation or energy-saving obligations due to their comparatively small scale. Signing voluntary agreements then is a highly recommendable way for local authorities to guide energy-saving and climate-friendly practices in the service sector.

Instrument

$$\text{Ratio} = \frac{\text{Enterprises which sign voluntary agreement}}{\text{Total service enterprises}}.$$

Data Source

Economic and Industrial Committee, Bureau of Commerce, Low-Carbon Workgroup, DRC.

Data Sheet

Parameter	Unit	Quantity
Number of companies which signed the volunteer agreement	Company	
Total number of companies	Company	

Key Responsible Entity for Action

Bureau of Commerce, Economic and Industrial Committee.

Action Checklist

Item	Yes	No	N/A
Controlled Items			
Specific institution is responsible for and monitoring the process of voluntary agreement system			
There are companies in the service sector that have signed the voluntary agreement			
A voluntary agreement template is available with clear terms			
General Items			
Activities organized to promote voluntary agreement signing			
Assessment conducted for the implementation of the agreements			
Optional Advanced Items			
Policy incentives such as subsidies/tax exemption to encourage service companies to sign voluntary agreement are available			

(3) Number of Companies in Energy Service, Energy Management and Contracting, CDM Services, Energy Consulting, etc.

Definition

This indicator is to measure the number of companies in the municipal service sector which provide energy management and contracting, energy consulting and services related to carbon emission mitigation.

Scope

Energy service includes various and extensive energy-related consulting services, such as energy-saving planning, services related to carbon emission mitigation in different areas, energy management and contracting, CDM services, energy consulting and other potential services.

Relevance

The prosperity of the local energy service business can reflect the development of the energy management capability in municipalities. This indicator aims to introduce and promote the advanced professional energy-saving concept and the methodologies of outsourcing energy management in local business.

Data Source

Economic and Industrial Committee, Bureau of Commerce, Low-Carbon Workgroup, websites of companies, local yellow pages.

Key Responsible Entity for Action

Economic and Industrial Committee.

Action Checklist

Item	Yes	No	N/A
Controlled Items			
Policy incentives available for registering and setting up energy service business			
General Items			
Propaganda activities available to promote energy service, such as workshops, exhibitions and commercials, etc.			
A vibrant, distinctive and regulated energy service market established including players such as professional energy service companies and integrated energy companies			
Optional Advanced Items			
Establish a relatively complete energy service system; professional energy service companies expanding service capabilities and fields; energy management contract turning into the main energy-saving business model			

3.2.3 *Low-Carbon Agriculture*

(1) Fertilizer Consumption (ton/ha)

Definition

Fertilizer consumption refers to the quantity of chemical fertilizers used (ton per hectare) in farms each year.

Scope

The fertilizer consumption in this indicator is subject to the *National Statistic Yearbook* (2010), Consumption of Chemical Fertilizers in Agriculture. It includes nitrogenous fertilizer, phosphate fertilizer, potash fertilizer and compound fertilizer applied in farms.

Farms refer to areas of land sown or transplanted with crops regardless of being in a cultivated area or non-cultivated area. Areas of land re-sown due to natural disasters are also included. At present, the sown area of crops are classified in the following nine categories: grain, cotton, oil-bearing crops, sugar crops, flax crops, tobacco, vegetables and melons, medicinal materials and others. For more details please refer to the *National Statistic Yearbook*.

Relevance

Agriculture is one of the main anthropogenic GHG sources. Nitrogenous fertilizer emits N_2O; the production of agrochemicals consumes a large quantity of energy and discharges GHGs. Currently nitrogenous fertilizer usage per acre in China is about two times the level in the USA. There is a large potential to achieve direct emission mitigation in agriculture.

Instrument

According to the *National Statistic Yearbook* (2010), the consumption of chemical fertilizers is calculated in terms of volume of effective components by means of converting the gross weight of the respective fertilizers into the weight containing the effective component (e.g. nitrogen content in nitrogenous fertilizer, phosphorous pentoxide contents in phosphate fertilizer, and potassium oxide contents in potash fertilizer). Compound fertilizer is converted in regard to its major components. The formula is:

Volume of effective component
  = Physical quantity
    × effective component of a certain chemical fertilizer (%).

Fertilizer consumption (ton/ha)

$$= \frac{\text{Fertilizer consumption in agriculture}}{\text{Sown area of crops}}.$$

Data Source

Agriculture Bureau, *Statistic Yearbook.*

Key Responsible Entity for Action

Agriculture Bureau.

Action Checklist

Item	Yes	No	N/A
Controlled Items			
Establish and implement statistical approaches for data collection on fertilizer consumption			
Efforts paid to reduce fertilizer usage, such as promoting more environmentally friendly fertilizer, introducing scientific integrated field management, etc.			
General Items			
Public awareness promotion activities carried out about scientific fertilizer application			
Pollution-free/organic agriculture products are promoted			
Optional Advanced Items			
Soil testing and proper fertilization used in all sown areas			
Developed local technical specifications regarding proper fertilization based on soil testing			

(2) Share of Rural Households with Biogas Digesters in Households Suitable for Building Biogas Digesters (%)

Definition

This indicator refers to the number of rural households which have established biogas digesters against the total number of rural households suitable for building biogas digesters.

Scope

Since natural and agricultural conditions vary greatly across China, the total number of rural households that are suitable for building biogas digesters can be defined based on investigations by the local Agriculture Bureau.

Relevance

This indicator reflects the application of renewable energies in rural households in a municipality. Converting biomass into methane as a renewable energy source for household cooking and lighting can reduce carbon emissions caused by agricultural residues. It can also act as an efficient substitute for traditional coal and firewood, not only helping to mitigate emission but also benefiting the local air quality.

Instrument

Local survey.

Data Source

Agriculture Bureau, Rural Energy Office.

Key Responsible Entity for Action

Agriculture Bureau, Rural Energy Office.

Action Checklist

Item	Yes	No	N/A
Controlled Items			
Assessment conducted to test applicability of biogas digesters in rural area			
Plan in place on expanding biogas application in rural area			
The established biogas digesters are mostly in operation after being built-up (above 90%)			
General Items			
Establish mechanism to resolve problems identified in operating biogas digester			
Technical specifications developed for biogas usage			
Optional Advanced Items			
Biogas plan incorporated into integrated RE utilization planning			

(3) Treatment and Reuse of Agriculture and Forestry Residues

Definition

This refers to how residues from agricultural and forestry production are treated to reduce carbon emission, environmental pollution and reused as resources.

Scope

Residues in agriculture and forestry include straws of various crops, barks and roots of trees, manure and so on.

The way of treatment and reuse includes all techniques which can help reduce pollution and carbon emission of the residues, e.g. through combined heating plant (CHP), household biogas digester, converting to methane, used as feedstuff, processed into construction materials and so on.

Relevance

Agricultural and forestry residues are one of the main environmental polluting sources in China's rural area. Traditional treatment methods such as burning and natural digesting will result in direct emission of CO_2 and CH_4. This indicator focuses more on the local living environment for rural residents, with the purpose of promoting cleaner and more efficient methods to replace the traditional treatments.

Instrument

Local surveys.

Data Sources

Agriculture Bureau, Rural Energy Office.

Key Responsible Entities for Action

Agriculture Bureau, Rural Energy Office.

Action Checklist

Item	Yes	No	N/A
Controlled Items			
Attention is paid to the issue and specific institutions are monitoring and managing agricultural and forestry residues			
Informational support provided to local agricultural and forestry producers such as handbooks, documents and notices, on how to recycle and reuse residues			
General Items			
Policy incentives are available to encourage environment-friendly innovations in treating and applying residues			
Optional Advanced Items			
Technical specifications developed for treating agricultural and forestry residues			

(4) Energy-Saving Measures Applied in Agricultural Production

Definition

Energy-saving measures in agricultural production refer to whether there are energy-saving measures, including conservation of electricity, diesel, petroleum, water, etc., implemented in agricultural production and how efficiently these measures function.

Scope

It refers to the energy consumption and energy-saving measures in the agricultural production process, not including rural residential energy consumption. The measures cover all aspects of the agricultural production process, including plowing, irrigating, reaping, threshing, primary process, etc. Energy-saving measures include the conservation of not only electricity and fuel, but also other resources and materials like water, films, chemicals and so on.

Relevance

Traditional farming practices are energy inefficient. Diesel and petroleum are massively used in irrigating and reaping. This indicator aims to facilitate energy-saving practices in agricultural production.

Instrument

Local surveys.

Data Sources

Agriculture Bureau, Rural Energy Office, data on diesel, electricity and petroleum consumed in agricultural production is available from the *Statistic Yearbook*.

Key Responsible Entities for Action

Agriculture Bureau, Rural Energy Office.

Action Checklist

Item	Yes	No	N/A
Controlled Items			
A plan is set by relevant agencies to enforce energy saving in agriculture			
General Items			
Good practices, techniques and methods on energy saving have been provided to agricultural producers via various media like TV programs, brochures, leaflets, training courses, demonstration classes, etc.			
Optional Advanced Items			
Energy-saving innovations get support from the local government			
Technical specifications developed for energy conservation in agricultural activities			

3.3 Green Building

3.3.1 *Planning*

(1) Application of Energy Consumption Statistical Tool for Different Types of Building

Definition

This indicator can help assess the statistical performance of building energy consumption. The inventory of building energy consumption includes statistical charts of building energy consumption by different building types and different energy types.

Scope

Building types cover all kinds of civil buildings, including residential buildings, large public building, office buildings, etc.

Energy types include electricity, heat, coal, gas, renewable energy, etc.

Relevance

The development of a statistical system for urban building energy consumption is the fundamental basis for assessing the emission level of different buildings. Currently, the statistical information on the existing building stock and building energy consumption at the city level is rather incomplete. So it is necessary to set up inventory tools for the statistical system.

Instrument

In December 2009, the MOHURD established the Civil Building Energy Consumption and Energy Saving Information Statistics System, which was authorized by the National Bureau of Statistics of China. This system set up a whole set of scientific and detailed institutions and methodologies for building energy consumption statistics. The inventory of building energy consumption can refer to rules of the Civil Building Energy Consumption and Energy Saving Information Statistics System.

Data Source

Construction administration department.

Action Checklist

Assessment on the statistical scale and depth shall be conducted by construction administration departments.

Item	Yes	No	N/A
Controlled Items			
Carry out systematic statistical work based on building energy consumption according to the Urban Civil Building Energy Consumption Statistics Information Comprehensive List			
General Items			
Provide energy consumption statistical information on large public buildings			

(*Continued*)

(Continued)

Item	Yes	No	N/A
Provide energy consumption information on governmental agencies			
Provide complete working plan of collecting building energy consumption statistical information (including all building types contained in the comprehensive list)			
Optional Advanced Items			
Collect civil building energy consumption statistical information for more than 3 years			
Compose and release statistical analysis report of building energy consumption periodically			
Statistical analysis report of building energy consumption should be referred to in order to guide future building administration			

(2) Green Building Action Plan Developed

Definition

An action plan is needed to promote widespread application of green building. The main focus here is the planning of promoting green building with particular emphasis on government actions.

Scope

This indicator assesses whether there are municipal plans, principles and action guidance for promoting green building, whether there are municipal policies, rules, regulations and concrete measures for local authorities to promote green building, and whether there are specific institutions established for promoting green building.

Relevance

Assesses the efforts of the local government in promoting green building.

Instrument

The assessment is a combination of qualitative indicators such as whether there are municipal plans, principles and actions guidance for promoting green building and whether there are specific institutions established to promote green building together with quantitative indicators such as the number of green building policies released.

Data Source

Construction administration department.

Action Checklist

Item	Yes	No	N/A
Controlled Items			
Implement national policies for promoting green building			
Assign designated staff to work on promoting green building			
General Items			
Establish mid- and long-term plan for promoting green building.			
Establish yearly action plan for implementation			
Optional Advanced Items			
Establish yearly quantitative targets of green buildings constructed			

(3) Green Building Action Plan Developed for New Urban Areas

Definition

New urban areas refer to the integrated and comprehensive development and construction activities in the areas outside existing urban areas based on the overall urban planning, including the construction of an economic and technological development zone, satellite towns and new industrial or mining areas. Planning of building energy consumption in new urban area development should require strict implementation of building energy conservation standards for new buildings, encourage EPC projects and set market access criteria based on clear function design.

Scope

This indicator looks into whether building energy consumption plans, actions or guidance are applied to new buildings in new urban areas.

Relevance

Establishes high standards and high starting level for green building for new urban area development.

Instrument

Whether the municipality has established building energy consumption plans, actions or guidance for new buildings in new urban areas.

Data Sources

Local DRC, Transportation and Construction Committee, and Planning Bureau.

Action Checklist

Item	Yes	No	N/A
Controlled Items			
Strictly implement building energy conservation standards in new buildings			
Establish administration and supervision agencies for new buildings in new urban areas			
General Items			
Establish guidance to implement building energy consumption plans, actions or guidelines in new urban areas			
Optional Advanced Items			
Set clear targets for amount of new buildings meeting green building standards			
Establish guidance for implementing new building energy consumption standards in new urban areas			

3.3.2 *Developing Management*

(1) Overall Performance of Developing Energy Management

Definition

This indicator reflects the maturity of the market for building energy management using the quantity of buildings applying energy performance contracting (EPC). EPC refers to the market mechanism of energy conservation business that reduces energy consumption, hence energy bills, and uses the money saved to pay for related project costs.

Scope

EPC projects applied to all types of buildings.

Relevance

This market mechanism has very promising market potential and should draw special attention.

Ever since the EPC mechanism was introduced in China at the end of the 1990s, the energy conservation industry rapidly grew and the number of professional energy conservation companies increased dramatically. The service scope covers industries, buildings, transportation facilities and public agencies. In 2009, there were about 502 energy conservation service companies in China with a total output of up to 58 billion RMB. Services offered by these companies can save 13.5 million tce each year, which has positive implications on promoting energy conservation, reducing energy consumption and increasing employment. However, EPC programs in China have not drawn enough attention. There are also many challenges, such as limited tax incentives, lack of financing channels, small development scale and unregulated market, which all impose obstacles in the way to further promote energy conservation. Based on these actualities, we may see that EPC project promotion has great implications on reducing GHG emissions and energy consumption.

Instrument

Assessment will be based on the quantity of EPC applications on buildings. To promote EPC applications the State Council issued the Guidelines to Promote EPC and Energy Service Industries (#SC[2010]25) based on a number of documents including the Energy Conservation Law, Decision by State Council on Energy Conservation (#SC[2006]28) and Notice by State Council on Issuing Comprehensive Work Scheme for Energy Conservation and Emission Reduction (#SC[2007]15).

Data Sources

Construction administration department, energy administration department, Housing and Urban & Rural Development Bureau and Planning Bureau.

Action Checklist

Item	Yes	No	N/A
General Items			
Set up energy consumption targets and indicators for different building types			
Evaluate the implementation of targets and indicators			
Optional Advanced Items			
Take measures for those buildings exceeding energy consumption standards and supervise the use of financial support and tax incentives			
Promote LCCC Low-Carbon Property Management Guidance for key energy-intensive buildings			

(2) Efficiency Improvement of Existing Buildings

Definition

The indicator can reflect the progress of building energy efficiency retrofits in existing buildings. The compliance rate of building energy efficiency (%) is the share of compliance building areas in existing buildings out of the total building area.

Scope

China has already set up a target of building energy saving of 65% based on the existing 50% standards. Some regions (such as the four central-government directly administered municipalities) will implement a higher standard of energy saving of 75%. Both residential buildings and public buildings need to conform to such standards.

Relevance

Building energy efficiency is the key to reducing building emission and also the key to green building. The compliance rate of building energy efficiency is used to assess the progress of urban building energy efficiency and is also the key indicator to assess the efforts of urban low-carbon development in the building sector.

Instrument

Assesses urban building energy efficiency progress based on the retrofit compliance areas of building energy efficiency in the total existing buildings.

Data Sources

Construction administration, Housing and Urban & Rural Development Bureau and Planning Bureau.

Data Sheet

Parameter	Unit	Quantity
Standard of energy conservation in existing buildings	%	
Compliance rate of existing buildings reaches the energy conservation standards	%	
Compliance rate target of existing buildings reaches the energy conservation standards in 2015	%	

Key Responsible Entity for Action

Housing and Urban and Rural Development Bureau.

Action Checklist

Item	Yes	No	N/A
Controlled Items			
Establish medium- or long-term energy conservation retrofit plan for existing buildings (5–10 years)			
Implement short-term retrofit projects for existing buildings			
Whether the target of last year was achieved			
General Items			
Provide economic incentives for energy conservation retrofits			
Consider energy conservation retrofit when other retrofit projects for existing buildings are planned			
Optional Advanced Items			
Standards of energy conservation reconstruction retrofit should aim to achieve or exceed current local building energy efficiency standards			

(3) Application of Renewable Energies in Buildings

Definition

This indicator reflects the development of renewable energies application in buildings. It refers to the share of renewable energy use in the total energy consumption in buildings (%).

Scope

Solar heating, solar PV, superficial geothermal energy.

Relevance

Renewable energies are basically clean energies, thus extending the application of renewable energies in buildings will greatly reduce the emission level in buildings and will be an effective way to control building emissions. In light of this, indicators that describe the application status of renewable energies in buildings are very important to assess municipal low-carbon development efforts in the building sector.

Instrument

Refer to information on building energy consumption collected based on the Civil Building Energy Consumption and Energy Saving Information Statistics System.

Data Source

Construction administration department.

Action Checklist

Item	Yes	No	N/A
Controlled Items			
Establish long-term renewable energy building application plan based on the available renewable energy resource, application potential and possibilities			
Establish short-term action guidance for renewable energy building application			
General Items			
Whether the target of renewable energy building development set last year can be achieved			
Provide policies to promote application of renewable technologies and products suitable for local buildings			
Establish assessment criteria for renewable energy application in buildings.			
Optional Advanced Items			
Provide economic incentives for renewable energy application in buildings			
Set requirements for renewable energy application and enforce during the approval process			
Set targets on the share of renewable energy consumption in total building energy consumption			

(4) Capacity Building for Green Building

Definition

This indicator tries to describe the social activities on promoting green building with special emphasis on citizens' participation.

Scope

This indicator looks into the number of dissemination activities to promote green building knowledge conducted, the number of training programs organized for green building professionals, the number of visits to green buildings and the number of patents of green building technologies, etc.

Relevance

This indicator assesses the level of understanding of green building from the social perspective as well as the government's efforts to increase public awareness of green building.

Instrument

Collect statistical information on the number of events dedicated to the dissemination of green building knowledge, training for green building professionals, visits to green building and the number of patents of green building technologies, etc.

Data Sources

Construction administration department, Planning Bureau, propaganda administration departments and related news offices/local television stations.

Action Checklist

Item	Yes	No	N/A
Controlled Items			
Plan developed for green building capacity building			
General Items			
Disseminate green building knowledge among decision makers in construction administrative departments			
Provide organized professional training for all green building professionals			

(*Continued*)

(Continued)

Item	Yes	No	N/A
Promote local market-based projects of 1-star and 2-star green building labeling			
Optional Advanced Items			
Provide financial support for such capacity-building events			

(5) Promoting Green Building (Narrow Definition)

Definition

This indicator reflects green building development by the total number of buildings meeting different green building standards.

For this research, green building standards refer to those outlined by the Assessment Standards of Green Buildings (GB/T 50378—2006) issued by the MOHURD. To be specific, green buildings refer to the buildings which can save resources (energy saving, land saving, water saving and material saving) to the maximum extent during their whole building life cycle, and can help protect the environment and reduce pollution, provide people with healthy, convenient and efficient space, and can harmoniously coexist with nature.

Scope

Green building evaluation labeling consists of green building design labeling (GBDL) and green building labeling (GBL). Green building can be graded from low to high in three ranks: one star, two stars and three stars. National green building innovation awards can be categorized as engineering awards and technologies and products awards. Engineering awards include comprehensive awards for green building innovation projects, special awards for intelligent buildings and energy-efficient building innovation. Technologies and product awards are granted to innovative and significantly effective new technologies, new products and new processes. There are three classes for the innovation prizes: first prize, second prize and third prize.

Relevance

For buildings to meet the standards of green building and receive awards in related areas, comprehensive evaluation of buildings is conducted, including that on energy saving, land saving, water saving and material saving, which can also show the environment protection and renewable energies application in buildings. So information on standards compliance and award winning can reflect a city's effort towards low-carbon development.

Instrument

China has already issued a series of policies and programs on green building standards and awards, including Management Approaches of Green Building Evaluation Mark Labeling, Notification on Promoting 1-Star and 2-Star Green Building Evaluation and Labeling, Detailed Rules on Green Building Evaluation Labeling Technologies, Additional Remarks on Detailed Rules on Green Building Evaluation Labeling Technologies, Management Approaches of National Green Building Innovation Awards, Detailed Rules on National Green Building Innovation Awards, and Appraisal Standards on National Green Building Innovation Awards. Here, indicators are assessed by collecting information on municipal buildings which meet the standards and win related awards.

Data Source

Construction Administration Department and Planning Bureau.

Action Checklist

Item	Yes	No	N/A
Controlled Items			
Set both annual and long-term green building targets			
General Items			
Number of green buildings increased compared with last year			
Standards to assess green buildings' local applications set			

(Continued)

(Continued)

Item	Yes	No	N/A
Optional Advanced Items			
There are green buildings winning innovation awards			
Promote green retrofit for existing buildings and set assessment local criteria for green building			

(6) Implementation of Incentives and Policy Instruments

Definition

Number of documents released in the areas of green building, building energy conservation and renewable energy application in buildings.

Scope

All policies and documents on promoting green building, building energy conservation and renewable energy application in buildings are included.

Relevance

Green building, building energy conservation and using renewable energies in buildings are the most popular ways for low-carbon building development. They have great impacts on building emission reduction.

Instrument

Assess the number of documents released in the areas of green building, building energy conservation and renewable energy application in buildings.

Data Sources

Construction administration department, energy administration department.

Key Responsible Entity for Action

Housing and Urban and Rural Construction Bureau.

Action Checklist

Item	Yes	No	N/A
Controlled Items			
Develop incentive policies and concrete implementation plans			
Application of major RE sources locally available is encouraged in the incentives			
General Items			
Assessment has been conducted regarding the approval and allocation of the incentives/allowances			
Establish local green building administration			
Optional Advanced Items			
Relevant results are integrated into local policies and standards			

(7) Demonstration Projects of Best Available Low-Carbon Technologies

Definition

Demonstration projects of the best available low-carbon technologies refer to those projects which combine low-carbon technologies and buildings.

Scope

Demonstration projects of the best available low-carbon technologies in all kinds of buildings are accounted.

Relevance

Commitment by the local government to promote demonstration projects of the best available low-carbon technologies will have positive impacts on building emission reduction.

Instrument

Assess the quantities of demonstration projects of the best available low-carbon technologies.

Data Source

Construction administration department, energy administration department.

Key Responsible Entity for Action

Housing and Urban and Rural Construction Bureau.

Action Checklist

Item	Yes	No	N/A
Controlled Items			
Develop plan for low-carbon building technologies, R&D and piloting			
General Items			
There are pilot programs where low-carbon technologies are applied to buildings			
Optional Advanced Items			
Pilot programs of low-carbon technologies for buildings showed effective demonstration effect			

3.4 Low-Carbon Transport

3.4.1 *Low-Carbon Transportation Strategy and Planning*

(1) Emissions Inventory for Different Transportation Modes Established

Definition

This indicator reflects the effort of a municipality to establish and improve the database which records detailed emissions of urban transportation. Inventory categories mainly include public land transportation and private transport (vehicle, electric bicycle, subway, etc.). Once the inventory is established, it should be updated annually.

Scope

Emissions in this indicator refer to the six main greenhouse gases listed by the Kyoto Protocol, i.e. CO_2, CH_4, N_2O, HFCs, PFCs and SF_6. Inventory may not be comprehensive and cover all six GHGs at first, but data on CO_2 emission is necessary. Different transportation modes include private vehicles, taxis, motorbikes, buses and urban rail transit.

Relevance

An inventory on transportation emissions is the basis to make a low-carbon transportation strategy and action plan.

Data Source

Transportation Bureau and Statistic Bureau.

Instrument

Emission = Number of motor vehicle × VKT × (E/VKT).

VKT means vehicle kilometers of travel; E/VKT means energy efficiency. (Georges Darido, Mariana Torres-Montoya and Shomik Mehndiratta, "Urban traffic and CO_2 emissions: Some characteristics of Chinese cities.")

Key Responsible Entity for Action

Safety Technology Department of Transportation Bureau.

Action Checklist

Item	Yes	No	N/A
Controlled Items			
Establish dedicated department in charge of inventory on transportation emissions			
There is an inventory tool in place			
General Items			
Inventory covers all transportation modes listed			
Inventory covers all GHG emissions			
Optional Advanced Items			
Emission database is updated regularly			

(2) Low-Carbon Transportation Strategy and Action Plan Developed

Definition

A transportation strategy is based on overall city planning, and involves consideration of the economy, climate, environment, etc. This indicator shows whether the concept of low-carbon development is comprehensively integrated in the transportation strategy and action plan.

Scope

The transportation strategy and action plan shows the roadmap to achieving the target in the transportation plan. The transportation strategy and action plan involves the following areas: road and rail construction planning, traffic policy, slow traffic, etc. Transportation strategies here include those for both rural and urban areas.

Relevance

The transportation strategy and action plan has a direct influence on a city's future development trend. Formulating a low-carbon transportation strategy and action plan can nudge future development in the low-carbon direction.

Data Sources

Transportation Bureau, Municipal DRC, Traffic Police Station, Highway Bureau, Planning Bureau.

Key Responsible Entity for Action

Transportation Bureau.

Action Checklist

Item	Yes	No	N/A
Controlled Items			
Design and update low-carbon transportation strategy and action plan regularly			
Concrete target of low-carbon transportation is set in strategy and action plan			
Responsible institutions for implementation are clearly illustrated in action plan			
General Items			
Research on low-carbon transportation with consideration for local context			
Integrate low-carbon transportation strategies into the city's master plan (e.g. its 12th Five-Year Plan)			
Optional Advanced Items			
Action plan is approved and integrated into a city's official plan (e.g. its 12th Five-Year Plan)			
Party responsible, timeline and budget are clearly illustrated in the action plan			

3.4.2 *Transportation Management*

(1) Integrated Transportation Management in Place

Definition

The target of integrated transportation management is to make traffic safe, convenient and energy saving through managing vehicles, people and roads in the transportation system appropriately. This indicator checks whether there are appropriate targets and a roadmap in the transportation management system to reach an ideally efficient traffic. Measures for mobility transportation include promotion of the concept, establishing coordination institutions (led by mayors and involving stakeholders), enhancing coordination tools (such as inclusive meetings, and enabling stakeholders to participate in transportation plan development, infrastructure construction and management).

Scope

Transportation management covers both rural and urban areas. Transportation management can use the following tools:

- Incentives: Subsidies for public transportation expenditure, road pricing, incentives for using slow traffic such as special parking lots and charging stations for e-bicycle at the work place, subsidies for bicycle rental, etc.
- Information: Establish information-sharing platforms and improve information service for use/interconnections of public transportation, etc.
- Regulations and standards: Development of parking lot standards, low speed zones, special bus lanes, vehicle travel limits based on plate number, lottery for private car plate numbers, etc.
- Infrastructure: Reduce parking lot capacity, BRT, improve street lighting, etc.

Relevance

One goal of integrated transportation management is to reduce fuel use while vehicles are idle, which is also a requirement of low-carbon transportation.

Data Sources

Transportation Bureau, Municipal DRC, traffic police stations.

Key Responsible Entity for Action

Transportation Bureau.

Action Checklist

Item	Yes	No	N/A
Controlled Items			
Funding and responsible departments are clearly defined for the implementation of strategies			
Assessment conducted on strategies and pilot projects based on the need and development trend of vehicles			

(Continued)

(*Continued*)

Item	Yes	No	N/A
General Items			
Revise strategy and pilot projects regularly			
Conduct assessment on both the quantity and quality of pilot projects implemented on integrated transportation management			
Optional Advanced Items			
Do existing strategies or pilot projects cover the four focus areas (incentive, information, regulation, standard and infrastructure)?			
Mechanism established to report and integrate assessment results into new planning activities			

(2) Average Commuting Time

Definition

This indicator reflects the general time spent on the way from home to workplace every day.

Scope

This indicator only includes time spent in urban traffic, while intercity transport time is not considered.

Relevance

Time spent in traffic can reflect the quality of transport planning and management. The more time is spent in traffic, the more fuel will be used. So the average commuting time also reflects the amount of transportation-associated carbon emissions. This indicator also reflects the operating efficiency of a city, for traffic jams are usually an important factor responsible for the low efficiency and bad mood of people.

Instrument

If data is not available, the average trip length for different means of transportation can be calculated according to a survey and then the total commuting time can be calculated.

Data Sources

Transport Bureau, Municipal DRC, traffic police stations, surveys, local research institutions.

Data Sheet

Parameter	Unit	Quantity
Average commuting time	Minutes	
Average commuting time in previous year	Minutes	

Key Responsible Entity for Action

Transport Bureau.

Action Checklist

Item	Yes	No	N/A
Controlled Items			
Research on average commuting time to work			
General Items			
Specific targets set to reduce commuting time			
Disclose the targets, strategies and action plan to the public			
Optional Advanced Items			
Check if target has been achieved			
Action plan integrated into official planning			

(3) Share of New Energy Vehicles in Public Vehicles (%)

Definition

This indicator is to reflect the progress of promoting new energy vehicles in a city.

Scope

New energy vehicles refer to automobiles using unconventional fuel as a power resource (or using conventional fuel in an unconventional power system). New energy vehicles include HEV, BEV (battery electrical vehicle), FCEV (fuel cell electrical vehicle), CNG, LNG, bio fuel vehicle, etc. (Management Guidelines for New Energy Vehicle Manufacturer and Production, issued by Ministry of Industries and Information on June 17, 2009).

Relevance

Generally, unconventional vehicle fuels are non-fossil or have a lower emission factor. Promoting new energy vehicles can reduce the total amount of carbon emissions and improve the energy mix associated with transportation. Increasing the share of new energy vehicles in public vehicles, buses and taxis shows a municipality's commitment to low-carbon transportation.

Data Sources

Transportation Bureau, Municipal DRC, traffic police stations, Public Property Management Center, surveys.

Data Sheet

Parameter	Unit	Quantity
Share of new energy vehicles among government vehicles	%	
Share of new energy vehicles among public vehicles	%	
Share of new energy vehicles among taxis	%	
Share of new energy vehicles among sanitation trucks	%	

Key Responsible Entity for Action

Public Property Management Center, Transportation Bureau.

Action Checklist

Item	Yes	No	N/A
Controlled Items			
Channels established to collect data on existing vehicle fuel type and efficiency			
Targets set for promoting new energy vehicles			

(Continued)

(Continued)

Item	Yes	No	N/A
Organize awareness-raising activities for knowledge of new energy vehicles for people responsible for procurement			
Set up incentives for new energy vehicles such as financial subsidies, tax reduction, etc.			
General Items			
Set up detailed policies to purchase new energy vehicles for public vehicles, taxis and municipal vehicles			
Set up pilot projects for new energy vehicles			
Build infrastructure for new energy vehicles			
Optional Advanced Items			
The number of new energy vehicles increased this year			
The share of new energy vehicles increased this year			

(4) Measures to Promote Efficient/Renewable Energy Bulbs for Street Lighting

Definition

This indicator includes two aspects: measures to promote efficient bulbs used in street lighting, and measures to promote renewable energy in street lighting. This indicator reflects efforts to reduce energy consumption in street lighting.

Scope

Street lighting refers to lighting devices installed on roads to provide necessary visibility to vehicles and pedestrians at night. Efficient lighting includes self-ballasted regular fluorescent lamps, three-color double-ended straight-tube fluorescent lamps (T8, T5) and metal halide lamps, high pressure sodium lamps and other electric light source products, semiconductor (LED) lighting products, as well as the necessary supporting ballasts. See Interim Measures on Management of Financial Subsidy for High Efficiency Lighting Products (#MOF-MOHURD[2007]1027).

Relevance

Since the street lighting devices are used outdoors, it is quite convenient for them to use various renewable energies, such as solar energy. In addition, energy consumption in street lighting is significant and the municipality can easily influence the purchase of street lighting devices.

Instrument

Percentage

$$= \frac{\text{Length of streets using efficient bulbs in street lighting}}{\text{Total length of all streets}}$$

Data Sources

Transportation Bureau, Municipal DRC, Street Lighting Management Office.

Key Responsible Entity for Action

Street Lighting Management Office.

Action Checklist

Item	Yes	No	N/A
Controlled Items			
Illustrate clearly who is responsible for maintaining the existing efficient bulbs and renewable resource devices			
Assessments are conducted regarding the quality and energy use of street lighting			
Set up pilot projects to promote efficient bulbs and renewable resource devices			
General Items			
Amount of energy saving in street lighting through efficient bulbs and renewable resource devices is calculated			
Optional Advanced Items			
Targets of using efficient bulbs and renewable resource devices in street lightning are achieved			

(5) Assessment of Public Transport Service

Definition

This indicator reflects the level of convenience a city's public transport system provides.

Scope

Public transport includes taxis, buses, subway, light rail, BRT (bus rapid transit), trams, etc. The aspects of assessment include accessibility, punctuality, comfort level, affordability, etc.

Relevance

Convenient public transport provides a strong incentive for people to choose public transport for mobility. Increasing numbers of people choosing public transport instead of private vehicles will lead to energy consumption reduction associated with transport.

Data Source

Transport Bureau, Municipal DRC, traffic police stations.

Data Sheet

Parameter	Units	Quantity
Percentage of people who are satisfied with punctuality of buses	%	
Number of buses in operation/10,000 residents	Standard unit	
Percentage of urban area with bus station within 300 m	%	
Average transfer time (referring to average time citizens spend transferring within transportation hubs)	Minutes	
Percentage of total investment in public transport infrastructure out of total transport investment	%	
Percentage of public transport financial subsidies in the total cost of bus operation	%	

Key Responsible Entity for Action

Bus companies.

Action Checklist

Item	Yes	No	N/A
Controlled Items			
Consider accessibility and convenience in station designs and planning			
General Items			
Develop guidelines or action plans that enhance priority for public transport development			
Establish stable and regulated bus subsidies			
Establish intelligent dispatch system			
Public transport assessment on accessibility, punctuality, comfort, affordability is done regularly			
Optional Advanced Items			
Activities to increase accessibility, punctuality, comfort, affordability			
Have urban public transport development funds			

(6) Number of Related Awareness-Raising Events Per Year

Definition

This indicator is to reflect the effect of increasing low-carbon awareness.

Scope

Low-carbon awareness refers to the ability to perceive, to feel or to be conscious of living and working in a low-carbon way. Awareness-raising events include public campaigns, seminars, workshops and other communications activities organized for the public.

Relevance

Low-carbon awareness-raising events will induce people to change to a low-carbon lifestyle. The change in consumer activities will help a city to achieve low-carbon development step by step.

Data Sources

Transport Bureau, Municipal DRC, traffic police stations, Publicity Department.

Data Sheet

Parameter	Unit	Quantity
Number of low-carbon awareness-raising events in the year	Number of times	

Key Responsible Entity for Action

Transport Bureau.

Action Checklist

Item	Yes	No	N/A
Controlled Items			
All residents have opportunities to participate in awareness-raising events			
General Items			
Organize low-carbon awareness-raising events regularly			
Optional Advanced Items			
Assessment of effect of events is done regularly			
The number of participants and number of events increased this year			

(7) Number of Persons Trained in Low-Carbon Capacity-Building Activities Per Year

Definition

This indicator reflects the results of capacity building for low-carbon transport.

Scope

Training and other capacity-building activities are offered to staff in charge of transport-related work in the municipality, including mayors, transport committee officials, transport bureau officers and professional entities. The content of low-carbon training and capacity building includes general knowledge of low-carbon development and low-carbon transport, new technologies, new energy, operation and maintenance of new equipment, etc.

Data Sources

Transportation Bureau, Municipal DRC, transportation committee.

Data Sheet

Parameter	Unit	Quantity
Number of people participated in training and other capacity-building activities in the year	People	

Key Responsible Entity for Action

Transport Bureau.

Action Checklist

Item	Yes	No	N/A
Controlled Items			
Organize activities for all relevant responsible staff			
Training content is linked to practical work and can be implemented in a short time			
General Items			
Hold events regularly			
Every event has a theme with specific targets			
Applicable training material is available			
Optional Advanced Items			
Assess outcomes of the events (such as coverage rate of stakeholder staff, etc.) regularly			
Training results such as the number of people trained or pilot projects initiated can be verified			

(8) Planning and Maintenance of Slow Traffic Infrastructure

Definition

This indicator reflects the results of promoting slow traffic system in a city.

Scope

Slow traffic refers to non-motorized transport, generally with a speed less than 15 km/h, including walking and cycling. Infrastructures for slow traffic refer to public facilities designed to provide convenience to pedestrians and bicycles, including sidewalks, overpasses, bicycle lanes, etc.

Relevance

The slow traffic system is one of the various modes of transportation. As walking and cycling result in zero-carbon emission, they are important components of low-carbon transportation.

Data Sources

Transportation Bureau, Municipal DRC, traffic police stations.

Key Responsible Entity for Action

Transportation Bureau.

Action Checklist

Item	Yes	No	N/A
Controlled Items			
Slow traffic planning integrated into master plan			
Check the quality of existing infrastructure for slow traffic			
General Items			
Assess the implementation of slow traffic planning.			

(*Continued*)

(*Continued*)

Item	Yes	No	N/A
Set up demonstration projects of slow traffic (e.g. vehicle-free walkways, parking areas only for bicycles)			
Action plan is made and implemented to maintain and improve the quality of existing slow traffic infrastructure			
Optional Advanced Items			
Bicycle renting facilities are provided in the city			
Have policies and funds to support bicycle renting			

Chapter Four

Assessment of Dezhou's Low-Carbon Development

4.1 Facts of Dezhou City

Dezhou is located in the lower reaches of the Yellow River in the northwest of Shandong Province. The jurisdiction of Dezhou includes one district (Dezhou District), two cities (Yucheng and Leling), eight counties (Ningjin, Qihe, Lingxian, Linyi, Pinyuan, Wucheng, Xiajin and Qingyun) and two economic development zones (Dezhou Economic Development Zone and Dezhou Canal Economic and Trade Development Zone). Dezhou covers a total area of 10,300 km² with a registered population of 5.7 million (2010).

Dezhou is blessed with a long history and rich cultural heritage. It is one of the birthplaces of Longshan culture. It has been a capital city for state government or county government ever since the Qin and Han dynasties. During the Ming and Qing dynasties it was one of the country's 33 largest business center cities. The Yellow River culture, Yan culture and Qilu culture had a long history here, and Dayu culture and Confucian culture are deeply rooted in Dezhou.

From the geographical point of view, Dezhou is located in the Bohai Bay economic circle and lies in the connection zone for two major economic zones in the east and in the north (Figure 4.1). Beijing-Hangzhou Grand Canal passes through the city. Dezhou is also the transport hub in the country, with many major railways (Beijing-Shanghai line, Dezhou-Shijiazhuang line, Jinan-Handan line, Beijing-Shanghai Speed Train and Dezhou-Yantai line under construction), three highways (Beijing-Fuzhou, Jizhou-Liaocheng,

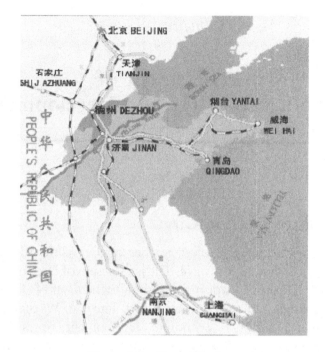

Figure 4.1   Map of Dezhou.

Qingdao-Yinchuan), five national roads and fourteen provincial roads passing through the city.

Dezhou is a typical Yellow River alluvial plain, an important agricultural production base for grains, vegetables and cotton in Shandong Province. The city has oil, coal, natural gas and other mineral resources too. The solar resource in Dezhou is 2,592 hours of annual average sunshine or 61%, with annual solar radiation of 124.8 kcal/cm². According to China's five categories for solar energy, Dezhou belongs to the third category — with a medium level of resource and economically feasible for development. However, there is no significant wind or water resource available for energy.

In recent years, socio-economic development in Dezhou has made remarkable achievements. In 2012 Dezhou's GDP topped 223.056 billion *yuan*, an increase of 12.1% over the previous year based on comparable prices. The primary sector contributed 24.439 billion *yuan*, an increase of 5.2%; the secondary sector 120.865 billion *yuan*, an increase

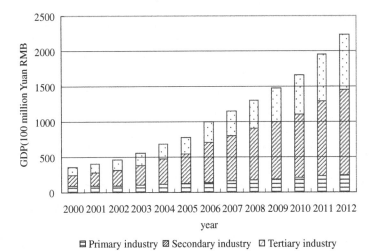

Figure 4.2   Ratio Between the Three Sectors in Dezhou (2000–2012).
Data source: Dezhou Yearbook 2012 and *Dezhou Statistical Release 2012.*

of 14.7%; and the tertiary sector 77.752 billion *yuan*, an increase of 10.2%. The ratio between the three sectors increased from the previous year's 11.8 : 54.3 : 33.9 to 10.9 : 54.2 : 34.9 this year, indicating a slight increase in the proportion of the tertiary sector (Figure 4.2).

4.2 Background

In the context of global climate change, many countries including the United Kingdom, those in the European Union and Japan have been developing corresponding strategies, setting mandatory targets, introducing legislations, regulations and standards, carbon taxes, fostering the emissions trading market to provide financial assistance, adjusting policies on government procurement, supporting R&D and technology transfer, etc., all with the common goal of developing low-carbon economies and low-carbon cities. These initiatives provide valuable experience for China to learn from in its own low-carbon economy development and low-carbon city construction.

The Chinese government has set explicit goals in the Twelfth Five-Year Plan of "green development and building a resource-saving and environment-friendly society", and that "in the face of increasing resource and environmental constraints, we must be aware of the environmental crisis, and establish a green, low-carbon development philosophy, focus on energy conservation, enhance incentives and regulations, accelerate the construction of resource-saving and environment-friendly production methods and consumption patterns, build capacity for sustainable development, and improve the level of ecological civilization". Faced with the dual challenges of economic transformation and "green revolution" both domestically and globally, China needs to follow the international trend of low-carbon economy and green development and turn pressure into opportunities.

As one of the developing countries actively responding to climate change, China initiated various programs for its low-carbon city development. At the national level, China identified five provinces and eight cities as low-carbon city pilot sites, where low-carbon development planning and supporting policies are developed, and low-carbon industries such as new energy and other strategically important industries are being fostered to reduce emissions and improve environmental protection. At the local level, cities are the main carrier of national economic development, and national low-carbon economic goals are ultimately allocated and to be achieved at the city level. Therefore low-carbon development is an important national initiative to address climate change, but also an important future direction for cities' development and transformation.

Dezhou City, although its economy is not huge, is at the critical stage of rapid development. It has the strong urge to pursue further economic development. In the context of low-carbon transition, Dezhou must change the mode of its economic growth, to explore low-carbon transition to ensure sustainable development for the future. The Twelfth Five-Year Plan period is a critical time for Dezhou City to change from its traditional high energy consumption growth to a sustainable low-carbon one. The municipality of Dezhou is committed to taking full advantage of its new energy industry and infrastructure resources, vigorously developing a low-carbon economy, building a low-carbon society and creating a low-carbon environment.

4.2.1 *An Industrial City*

Dezhou shows strong features of an industrial-based economy, although the service sector has grown rapidly in recent years and agriculture has also witnessed a steady increase. During the Eleventh Five-Year Plan period, Dezhou experienced a rapid growth of its industrial economy. The proportion of industries in the city's economy continues to increase. By 2010 the number of industrial enterprises above the designated size reached 3,477, the industrial added value exceeded 100 billion *yuan* for the first time, total industrial assets exceeded 200 billion *yuan*, and the main business revenue exceeded 400 billion *yuan*. Compared to 2005, Dezhou's industrial added value, the main business income and profits in 2010 increased by 1.87 times, 2.17 times and 1.75 times respectively. All these indicators achieved an average annual growth rate of more than 20% which exceeded their targets for the Eleventh Five-Year Plan period. At the same time, priority industries showed healthy development. Thirty-five industries completed continuous year-on-year growth in total added value. Fourteen of them showed a higher growth rate than the average rate of the city. Nine industries including communications equipment and other electronic equipment manufacturing, special equipment manufacturing and pharmaceutical manufacturing increased by over 25% — a rate much higher than the city's average of 16.42%. In 2011, the number of industrial enterprises above the designated size in Dezhou was 3,116, with a total added value of 100.135 billion *yuan*, accounting for 51.3% of the GDP.

For the Twelfth Five-Year Plan period, Dezhou plans to pursue further industrial economic development through the scientific approach to development, pursuing both quality and quantity of growth, transforming and upgrading traditional industries while developing new high-tech industries, fostering both technological innovation and industrial upgrading as well as large joint ventures, and targeting leap-forward industrial development via accelerating the construction of basic industries, relying on manufacturing and high-tech industries as the leader of the new industrial system, and cultivating large enterprises. Its industrial development strategy is set to be "pursuing structural adjustment and ensuring growth".

4.2.2   *Strong Urge for Development*

As an industrial city, Dezhou's economic development relies on traditional industries, such as textiles, agricultural product processing, chemical manufacturing, and electricity and heat. In the rapid economic take-off in the Shandong Peninsula, Dezhou's pace is rather slow compared with its peer cities because it is located in the inner part of the province. In Dezhou the disposable income of urban residents is lower than the average of the Shangdong people, and its GDP ranks eleventh among Shandong's 17 prefecture-level cities (see Figure 4.3).

In 2011, urban residents' per capita disposable income in Dezhou was 19,770.80 *yuan*, 2,039.2 *yuan* less than the national average of 21,810 *yuan* for the same period; urban residents' per capita consumption expenditure was 12,700.47 *yuan*, 2,460.53 *yuan* less than the national average; rural per capita net income was 8,350 *yuan*, 1,373 *yuan* higher than the national average of 6,977 *yuan* (see Tables 4.1 and 4.2). In 2011 Dezhou's urban per capita disposable income was 3,021 *yuan* lower than the provincial average, ranking the second lowest among all 17 prefecture-level cities in Shandong Province, only

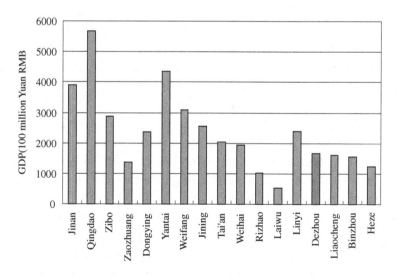

Figure 4.3   GDP of Prefecture-Level Cities in Shandong Province (2011).
Data source: Shandong Yearbook 2012.

Table 4.1   Major Economic Indicators in Shandong Province (2011).

Region	Total Household Income	Disposable Income	Consumption Expenditure	Total Household Expenditure
National average	23,979.2	21,810	20,365.71	15,161
The average in Shandong	24,889.80	22,791.84	19,341.27	14,560.67
Jinan	31,895.11	28,891.97	23,156.33	18,045.58
Qingdao	31,491.68	28,567.49	25,460.59	19,296.56
Zibo	26,541.23	24,955.40	20,142.51	15,994.09
Zaozhuang	23,082.94	20,193.38	19,805.73	13,463.09
Dongying	30,929.74	27,342.65	23,686.81	17,531.80
Yantai	28,815.01	26,541.75	22,942.62	18,395.10
Weifang	24,072.24	22,508.15	19,621.63	15,169.64
Jining	24,514.08	22,405.54	22,271.69	14,691.90
Tai'an	24,413.38	22,687.14	18,255.18	14,888.02
Weihai	27,750.35	25,290.24	22,456.95	17,001.88
Rizhao	21,288.18	20,097.91	19,604.46	13,781.37
Laiwu	27,484.30	23,508.87	20,639.64	14,218.74
Linyi	25,840.30	24,231.56	18,598.85	13,880.61
Dezhou	**21,903.87**	**19,770.80**	**19,101.95**	**12,700.47**
Liaocheng	21,768.95	20,649.24	17,574.94	13,846.94
Binzhou	24,912.86	22,540.34	20,978.41	14,808.24
Heze	18,501.36	16,658.09	14,869.84	11,215.99

Data source: Shandong Yearbook 2012.

higher than Heze City. Its rural resident net income is only higher than that in Liaocheng, Linyi and Heze among the 17 cities.

In summary, Dezhou is one of the slowest growing cities in light of the economic take-off in the Shandong Peninsula compared to its peer cities. Dezhou is eager to catch up with its peer cities and achieve rapid economic development, while it also has the responsibility to transform its traditional industrial-based economy into a low-carbon one. This poses a dual challenge for Dezhou.

Table 4.2 Rural Residents' Income and Expenditure in Shandong (2011).

Region	Annual Net Income	Net Income – Wage	Net Income – Household Operation	Net Income – Property	Metastatic Net Income	Net Income in Cash	Annual Total Expenditure	Expenditure – Household Operation
Jinan	10,412	4,971	4,399	470	571	9,683	9,548	2,747
Qingdao	12,370	5,418	6,122	281	549	11,195	13,192	4,222
Zibo	10,878	6,405	3,517	290	665	10,424	9,504	2,028
Zaozhuang	8,397	4,289	3,636	79	392	7,733	7,973	1,984
Dongying	10,025	3,732	5,430	388	476	8,962	11,639	4,606
Yantai	11,716	4,963	5,992	306	455	10,217	9,606	2,297
Weifang	10,409	4,550	5,193	267	399	9,669	11,116	3,786
Jining	8,712	4,223	3,758	369	363	8,036	7,774	2,182
Tai'an	8,974	4,677	3,800	163	333	8,198	6,877	1,460
Weihai	12,334	6,036	5,274	352	672	10,320	11,197	3,320
Rizhao	8,756	4,025	4,365	160	206	7,780	8,084	3,174
Laiwu	9,626	4,601	4,410	120	495	9,192	8,524	1,992
Linyi	8,018	3,102	4,305	200	411	7,103	7,949	2,628
Dezhou	**8,350**	**3,715**	**4,302**	**87**	**246**	**6,850**	**6,769**	**2,172**
Liaocheng	7,735	3,310	4,070	132	223	6,901	8,321	2,961
Binzhou	8,744	4,067	3,995	295	388	7,911	11,115	3,942
Heze	7,119	3,422	3,318	116	263	6,036	6,147	1,560

Data source: Shandong Yearbook 2012.

4.3 Practice and Highlights

In view of the city's industrial structure and natural resources, the municipality of Dezhou recognizes that to achieve leapfrog development it is necessary to accelerate low-carbon economy development and achieve economic transformation. The goal set is to "develop humanity-oriented eco-economy and build a harmonious Dezhou" and the strategy is identified as specifically accelerating the formation of a low-carbon industrial structure and transforming growth and consumption patterns. The political will for low-carbon city development is very strong. Dezhou has explored a variety of ways to develop a low-carbon economy.

First, Dezhou CPC Committee and the municipal government attaches great importance to low-carbon work and released its low-carbon development plan. As early as in 2007, Dezhou proposed to "develop humanity-oriented eco-economy and build a harmonious Dezhou". It clarified the mission of "starting with ambitious goals, construction with high quality, and aiming at leaps and bounds". The strategy is set to use "development of eco-economy and building a low-carbon Dezhou" as the guiding goal to vigorously promote its low-carbon city construction. Dezhou has organized several meetings to discuss the importance of low-carbon economy development. A low-carbon economy development plan was prepared and the Low-Carbon Economy Promotion Committee was established. The municipal government published Guidelines to Promote Low-Carbon Economy Development (Defa[2010]12) on May 10, 2010, setting low-carbon economy on a track of regulated and healthy development. This document sets the guiding ideology, basic principles and development goals, proposing the transformation of the economic development pattern, adjusting the economic structure, reducing the carbon intensity per unit of GDP, taking the leading role for the new energy industry and making Dezhou an example of a first-class low-carbon economy in China.

In June 2010, Dezhou was selected as one of the first four pilot cities for the Swiss-China Low-Carbon City demonstration project (LCCC) during the Action on Climate Change — Cooperation and

Dialogue with Swiss Cities because of pioneering work done on the solar energy industry and energy management. The Swiss Agency for Development and Cooperation has highly appraised and actively supported Dezhou to develop solar and other renewable energy industries, to promote the energy management system, to pursue building energy efficiency and to induce low-carbon lifestyles.

In 2011, Dezhou DRC organized the compilation of Dezhou's Twelfth Five-Year Plan for Low-Carbon Development, which analyzed Dezhou's opportunities and challenges based on identifying key areas for low-carbon development and assessment of the status quo. It is dedicated to seizing the opportunity to solve the short- and medium-term challenges faced, with an emphasis on the transformation of economic development patterns and adjustment of the economic structure. The Plan focuses on four aspects: renewable energy development, low-carbon production, low-carbon consumption and low-carbon management, with special attention on the initiation of a number of large low-carbon projects and the management of entangled systems in the low-carbon development. Its goal is to build solar and other new energy industries featuring low-carbon, clean production, to lead rational and conservative consumption trends, to build a complete and information-based carbon management system.

Second, Dezhou vigorously promotes the development of a renewable energy industry, making it the new economic growth driving force for the city. Dezhou is the world's largest base for the solar thermal industry, the base of new energy for the national Torch Program, and a demonstration city for renewable energy. In 2005, Dezhou proposed its China Sun City strategy. It has promulgated two policies: Accelerating the Implementation of China Sun City Strategy, and Promoting the Application of Solar Energy in Buildings. It has also prepared the New Energy Industry Development Plan and Renewable Energy Applications Development Plan (2009–2015). On August 31, 2010, the Dezhou Finance Bureau and the Housing and Urban Construction Bureau jointly issued the Dezhou Renewable Energy Demonstration City Project Fund Management Guidelines (Document #: DeCaiJian[2010]31). Through policy support and other measures, Dezhou has created a distinctive solar-based industry cluster, with leading enterprises including Himin, Yijianeng, Guoqiang, Zhongli, etc. By 2010, Dezhou had 126 companies in the solar energy and related business, with a total revenue of 8.2 billion *yuan* out of a total output value of 31.2 billion *yuan* of all new energy companies. At present, led by Himin Group, Dezhou has become the country's largest base for solar thermal research and development, production of solar water heaters, solar photovoltaics, energy-efficient building technologies and other related aspects.

For promotion of solar energy applications, the Dezhou government introduced two policies: the Solar for One Million Roofs Program in 2006 and the 5555 Program in 2008 that promotes solar photovoltaic applications. As of 2012, solar traffic lights were installed at 80 crossroads; solar light lamps were installed along 50 roads, 15 key scenic spots and some residential blocks; more than 12,000 solar light lamps were used; 137 km of roads were lit by solar lamps, out of which 100 km were major roads. Ninety-five percent of the buildings in the downtown and 50% of the buildings in all urban areas have solar energy application. In the central urban area of Dezhou, the total solar collector area reached over 600,000 m^2, the total capacity of solar photovoltaic applications was close to 50 MW, all these forming a unique Sun City landscape. In 2009 the Ministry of Finance (MOF) and Ministry of Housing, Urban and Rural Development (MOHURD) approved Dezhou as one of the first batch of national renewable energy demonstration cities and allocated a subsidy of 60 million *yuan* to Dezhou. In 2010 the Dezhou Economic Development Zone was selected to be among the first batch of concentrated photovoltaic demonstration areas, with a total installed capacity of 50 MW. In 2011, the National Energy Administration approved Dezhou as a state-level new energy demonstration city. In August 2011, Linyi County was selected as one of the national renewable energy demonstration counties and received 18 million *yuan* from the central government as financial assistance. Solar photovoltaic power generation plants are already in operation in a few places including Jinzijing residential block, Dezhou Museum and the Municipal Administration Center. Solar generation projects in Himin Group Weipai Tower, Himin Engineering and Technology Institute, New Lake Residential

Block have been selected as national demonstration projects and received grant funds of 33.65 million *yuan* from the central government.

As the country's leading solar energy industry base, Dezhou is actively involved in international cooperation and exchanges in the field of renewable energy. As early as September 16, 2005, Dezhou was awarded the Sun City title, becoming China's first city winning this award. In September 2010, Dezhou successfully hosted the Fourth World Conference of Sun Cities with "Solar energy changes life" as the theme, demonstrating the new solar-based energy industry and the latest achievements in related fields. At this conference, the MOHURD and MOF jointly named Dezhou a "renewable energy demonstration city". The conference promulgated the Dezhou Declaration — an important document that guides international solar cities movement. Besides, awareness of environmental protection, low-carbon and sustainable development was further raised.

Third, Dezhou has been endeavoring to implement a comprehensive energy management system and encouraging enterprises to strengthen their energy conservation management. In 2006 the NDRC organized the Energy Conservation among 1,000 Large Enterprises Program. Dezhou initiated energy audit programs for its key energy consumption enterprises and proposed to establish an energy management system.

In 2007, Dezhou was the first city to propose to conduct energy management system research in Shandong Province, and composed the only local energy management system standards: Energy Management System Requirements (DB37/T 1013-2008). Later it was revised to be more applicable in industrial enterprises: Industrial Energy Management System Requirements (DB37/T 1013-2009). Moreover, a supporting policy was also developed, including Industrial Energy Management System Implementation Guide (DB37/T 1567-2010), Steel Industry Energy Management System Implementation Guide, Industrial Energy Management System Audit and Evaluation Guide and five other local standards for an energy management system in order to better guide enterprises to develop an energy management system. Dezhou was fully involved in the two batches of pilot projects on an energy management system in Shandong Province, and provided on-site advice and training for 26 energy management system pilot enterprises in its six peer cities including Zibo, Dongying, Linyi, Liaocheng and Rizhao. Dezhou conducted evaluation of the energy management system construction of the first batch of four pilot enterprises. It published an Industrial Enterprise Energy Management System for energy management system professionals. All these efforts and inputs effectively helped to promote building an energy management system in Shandong Province.

Based on Dezhou's Key Energy-Using Companies Energy Management System Construction Implementation Plan, the municipality has officially launched the first batch of 16 companies to build an energy management system. By the end of 2013 all key energy-using companies with a consumption of over 10,000 tons will establish energy management systems. Given Dezhou's pioneering work and extensive experience in building energy management systems, Dezhou's Energy Monitoring Commission was selected by the Swiss Agency for Development and Cooperation as a consultant on energy management system application and promotion for other LCCC pilot cities in 2011. In addition, Dezhou's Energy Monitoring Commission was also selected by the NDRC as an authorized training institution for the 10,000 enterprises obliged to reduce their energy consumption during the Twelfth Five-Year Plan period.

4.4 Status Quo and Efforts

Dezhou is one of the pilot cities in the Sino-Swiss partnership project, the Low-Carbon City in China project (LCCC). Based on the LCCC Primary Indicator System and related logical framework, the report evaluates Dezhou's low-carbon development from two perspectives: status and level of effort. Major data sources include: (1) a variety of publicly available statistics, government documents, statistical bulletins; (2) field research and interviews with relevant departments and stakeholders; (3) domestic and international relevant low-carbon city index system research results.

The LCCC Primary Indicator System includes 15 indicators in 5 categories: economy, energy, infrastructure, environment and society. Among them, the economic indicators reflect a city's low-carbon economic development stage and implementation of the country's overall energy reduction targets; low-carbon energy indicators reflect regional low-carbon energy resources as well as the city's effort in upgrading its energy mix; infrastructure indicators show the urban infrastructure level and low-carbon consumption levels. Environmental indicators assess a city's green development. Indicators on society evaluate consumption patterns and social equity. Specific evaluation results are shown in Table 4.3.

4.4.1 *Low-Carbon Economy*

In accordance with evaluation by the LCCC Primary Indicator System, Dezhou's carbon productivity in 2010 was 0.32 in terms of 10,000 *yuan* GDP/tCO_2, an increase of 31.8% over 2005. Compared to the national average of 0.41, it is 22.7% lower. This shows that carbon productivity in Dezhou is rather low, and has large room for improvement.

During 2005–2010, energy consumption in Dezhou increased year on year, with energy consumption per unit of GDP showing a gradually declining trend — a decreasing rate of 5.1% per year and a total decline of 23.1%, a higher rate than the national average level of 19.1%. This shows its energy consumption was declining relative to its economic development, indicating an improvement. However, in terms of absolute energy consumption per unit of GDP, Dezhou's data was 1.196 tce/10,000 *yuan* GDP in 2010, still 15.8% higher than the national average (see Figure 4.4).

The key to Dezhou's low-carbon development is to reduce energy consumption per unit GDP and upgrade the energy consumption structure. The assessment shows that Dezhou is lagging behind the average level of Shandong Province and of China in terms of energy consumption per unit GDP, indicating Dezhou has a long way to go to fulfill its energy-saving tasks.

Table 4.3 Low-Carbon Development Assessment of Dezhou.

Nos.	Indicators	Unit	Dezhou		China		Effort (%)
			2005	2010	2005	2010	
	Economy						
(1)	Carbon productivity	10,000 *yuan*/ton CO$_2$	0.242	0.319	0.325	0.413	97
(2)	Energy intensity	tce/10,000 *yuan*	1.555	1.196	1.276	1.033	100
(3)	Decoupling index	—	0.76		0.85		80
	Energy						
(4)	Non-fossil energy in primary energy consumption	%	1.83	2.26	6.8	8.6	68
(5)	Per capita non-commercial renewable energy use	kgce/per capita	97.6	148.3	16.5	26.7	100
(6)	Carbon intensity of energy	ton CO$_2$/tce	2.648	2.615	2.405	2.341	75
	Infras.						
(7)	Energy consumption per unit building area for public buildings	kgce/m².	36.3	34.26	44 (large public buildings 2004)		80
(8)	Energy consumption per building area for residential buildings	kgce/m².	33.2	30.7	—		80
(9)	Ratio of green transport	Standard units	3.4	5.7	8.6	9.7	60

(*Continued*)

Table 4.3   (*Continued*)

Nos.	Indicators	Unit	Dezhou 2005	Dezhou 2010	China 2005	China 2010	Effort (%)
	Environment						
(10)	Percentage of days with API less than 100	%	83	87	51.9	81.7	80
(11)	Forest coverage rate	%		28.9	18.2	20.36	100
(12)	Domestic water consumption per capita per day	kg	113.83	90.79	149.80	105.04	83
	Society						
(13)	Urban-rural income ratio	%	2.38	2.48	3.22	3.22	100
(14)	Per capita CO_2 emission	ton CO_2/per capita	6.20	9.00	4.34	5.67	50
(15)	Low-carbon management institution	—	None	70%	None	100%	70

Note: For standards for capacity assessment see *LCCC Indicator System Methodology Report* (2012).

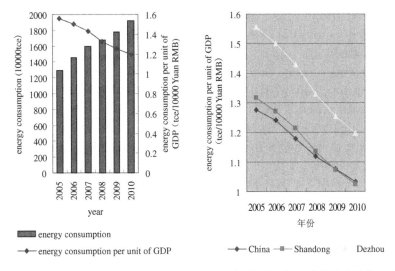

Figure 4.4   Unit GDP Energy Consumption in Dezhou (2005–2010).
Data source: Shandong Yearbook 2011, Dezhou Annual Statistical Release.

During 2005–2010, the country's carbon decoupling index was 0.86, while it was 0.85 in Dezhou, both showing a relative decoupling trend.

4.4.2 *Low-Carbon Energy*

Dezhou has a coal-dominated energy mix. In 2005–2010, Dezhou's share of non-fossil fuels in primary energy consumption rose from 1.83% up to 2.26%, but still lagged far behind compared with the 2010 national average of 8.6%. Dezhou's Twelfth Five-Year Energy Development Plan put forward the 2015 target of increasing the share of non-fossil energy consumption up to 12%. To achieve this ambitious goal, solar and other non-fossil energy sources need to see rapid development. The energy self-sufficiency rate in Dezhou is only 2.6%, i.e. the city has only 581,100 tons of raw coal production and 60,000 tons of coal products, while all the rest relies on external energy transferred. Data on the Shandong power grid shows that 99% is from thermal generation. Due to resource constraints Dezhou relies on coal, oil and natural gas. In recent years, Dezhou vigorously promoted the development of solar energy, biomass and municipal waste generation, but compared to the intended target, the challenge is still immense.

In terms of non-commercial renewable energy consumption per capita, during 2005–2010, Dezhou went from 97.6 kgce up to 148.3 kgce, an increase of 53.3%. At the same period the national average went from 16.5 kgce to 26.7 kgce, an increase of 61.82%. Dezhou is apparently doing better than the average level of China. See Table 4.4 for more details.

During 2005–2010, Dezhou's carbon intensity of energy dropped from 2.648 tCO_2/tce down to 2.615 tCO_2/tce. As Dezhou's proportion of coal consumption was significantly higher than the national average, its carbon intensity of energy consumption is 11.7% higher than the national average of 2.34 tCO_2/tce. This has to do with its fossil fuel-based energy mix and low energy efficiency.

4.4.3 *Low-Carbon Infrastructure*

In 2010, Dezhou had a total of 138 office buildings with an area of 762,300 m²; 20 large public buildings. In 2009, per unit area energy consumption in Dezhou's government office buildings was 30.15 kgce/m²•year, with 37.8 kgce/m²•year for large public buildings, and 34.26 kgce/m²•year as the weighted average. In 2004, China's average energy consumption per unit area for large public buildings was 44 kgce/m²•year.

In accordance with the requirements of the MOHURD, Dezhou has a rather complete statistics collection for the energy consumption of government office buildings and large public buildings, but the statistics for residential buildings are not yet complete. In 2005, Dezhou's urban resident per capita living area was 22.83 m². Total natural gas consumption in urban built-up areas was 25.47 million m³. Total consumption of liquefied petroleum gas was 4,081 tons. The district heating area in the urban city was 7.98 million m² of central heating. Non-agricultural population was 145.42 million. In 2010, Dezhou's urban resident per capita living area reached 31.87 m². Total natural gas consumption in urban built-up areas was 116.36

Table 4.4 Part of Non-commercial Renewable Energy Consumption in Dezhou.

Energy Type	Consumption	Source	Program	Conversion	Consumption in tce
Off-grid PV generation	5555 solar lighting program, nearly 5,000 solar lamps installed	Dezhou Plan for RE Application in Buildings	5555 program	In total 340 kW, saving about 1.24 million kWh annually	152.4
Off-grid PV generation	5,000 solar lamps installed for public areas and buildings	Same as above		In total 300 kW, saving about 1.2 million kWh annually	147.5
Solar heating	>90% downtown buildings and 50% urban buildings applied solar energy	Same as above		Solar collector per m² saves 216 gce	41018.4
Solar heating	>300 villages built solar shower rooms	Same as above	Solar Shower Room Program		10,000
Solar heating	31 hotels and schools installed solar collectors	China Sun City — Low-Carbon Dezhou	Solar Collector Program	Daily hot water supply capacity of 1,050 tons, saving 2,000 tce	2,000
Biomass	19 large biogas tanks built; 110,000 rural biogas tanks built	Dezhou Rural Biogas Plan (2010–2015)	Biogas Program	One rural tank saves 3 tons of firewood, one large tank saves 7,000 tons	211,213
Geothermal	>300,000 m² application targeted. 150,000 in 2005	Dezhou Plan for RE Application in Buildings	Geothermal Heat Pump		2842.3

million m^3. Total consumption of liquefied petroleum gas was 2,930 tons. The district heating area in the urban city was 13.8 million m^2 of central heating. Non-agricultural population was 168.31 million. Based on these data it is estimated that the residential building energy consumption per unit area in Dezhou decreased from 33.2 kgce/m^2•year in 2005 to 30.7 kgce/m^2•year in 2010.

During 2005–2010, the number of buses owned per 10,000 people increased from 3.4 to 5.7 standard units, while nationwide during the same period it increased from 8.6 to 9.7 standard units. Dezhou's figure is lower than the national average and far below the national requirement of 10–12.5 standard units/10,000 people given by the Ministry of Construction in the Urban Traffic Planning and Design Codes. According to the Guidelines To Development of Urban Public Transport With Priority by the Ministry of Construction (#MOC-UC[2004]38), it is required that urban public transport should accommodate 30% of total travel needs. This rate in Dezhou was only 2.7% in 2008, far below the national target. The urban area in Dezhou is not large, and cycling and walking accounted for 43.2% and 21.9% of all travels respectively. Together the two accounted for 65%. It is apparent that non-motorized transport makes up the bulk of transportation in Dezhou.

4.4.4 *Low-carbon Environment*

In 2005–2010, the percentage of days with an API of less than 100 increased in Dezhou, from 83% to 87% in 2010. In the same period nationwide it increased from 51.9% to 81.7%. In this sense Dezhou performed better than the national average.

In 2005–2010, the national per capita domestic water consumption decreased from 149.8 kg/person in 2005 down to 104.04 kg/person in 2010. In Dezhou it decreased from 113.83 kg/person in 2005 down to 76.43 kg/person in 2009, but rebounded to 90.79 kg/person in 2010.

Dezhou's forest coverage rate was 28.9% in 2010, higher than the national average of 21.36%, also exceeding the national requirement of 25% for northern cities. In 2005–2010, Dezhou's urban per capita green space showed an increasing trend. Dezhou initiated the Provincial Garden City Program and implemented greening projects for five consecutive years with nearly 200 million *yuan*, including constructing the New River Cultural Square, expanding 15 large parks (New Lake Scenic Area, Fairview River Scenic Area, etc.), greening 25 primary and secondary roads and 30 alleys in the downtown. Meanwhile, the government organized a program of adopting trees for enterprises, institutions and the general public. In 2008, 20,000 m^2 of green space and over 10,000 trees were adopted. Since 2005, the net increase of over 9 million trees and 1.80 million m^2 of green area has been more than the sum of the previous 10 years, forming a three-dimensional urban green system including scenic parks, green urban road network and suburban ecological green screens.

4.4.5 Low-Carbon Society

Dezhou's urban-rural income ratio during 2005–2010 was in the range of 2.38–2.45, lower than the national average of 3.22 over the same period. This shows that the income inequality in Dezhou is smaller.

Dezhou's CO_2 emission per capita increased from 6.2 tCO_2/person in 2005 to 9.0 tCO_2/person in 2010. During the same period, the national average went from 4.34 to 5.67. Dezhou's per capita CO_2 emission is 41.3% higher than the national average.

In terms of low-carbon management system, China has set explicit goals to achieve low-carbon transition. Pilot projects are conducted and the policy framework is being continuously improved. Since Dezhou developed its solar energy strategy in 2005, the municipality composed various low-carbon-oriented policies including the New Energy Industry Development Plan and Renewable Energy Applications Development Planning in 2010, the Guidelines on Accelerating Low-Carbon Economy Development (#Defa [2010]12) which set overall targets for 2015, and Dezhou's Twelfth Five-Year Plan Low-Carbon Economy Development which set further detailed targets for 2015. To accelerate low-carbon development, the city government established in June 2010 a Low-Carbon Development

Leading Group led by the mayor. Its secretariat is based in Dezhou DRC and is in charge of coordinating daily work, including researching and developing low-carbon development policies, following up on issues agreed, and implementing tasks assigned by the Leading Group. In addition, Dezhou is the pioneer in energy management system applications of the country. Dezhou has not developed any strategic plans for climate change adaptation yet, as China started this work not long ago.

4.5 Low-Carbon City Construction and Management in Dezhou

This section of the report gives an overview of Dezhou's low-carbon city construction and management based on the LCCC supporting indicators and their action checklists.

4.5.1 *Urban Management*

Back in early 2007, Dezhou clarified the mission it had tasked itself with, namely, "starting with ambitious goals, construction with high quality, and aiming at leaps and bounds". The strategy is set to use "development of eco-economy and building a low-carbon Dezhou" as the guiding goal, to vigorously promote low-carbon city construction. Dezhou established a number of institutions with clear delineation of their responsibilities, including the Low-Carbon Economy Promotion Committee. The municipal government established the Energy Office, Solar Office, Building Envelop Retrofit Office, Energy Monitoring Committee and Public Building Energy Management Agency. On May 10, 2007, the municipal government issued the Guidelines to Promote Low-Carbon Economy Development (#Defa [2010]12), setting low-carbon economy on a track of regulated and healthy development.

While developing its Twelfth Five-Year Plan, Dezhou had taken into account low-carbon development requirements. There is a special section on "fostering a low-carbon industrial system" in the plan, explaining   low-carbon   development   concepts,   low-carbon development models, low-carbon industry, low-carbon technologies, low-carbon lifestyle, etc. The plan also analyzes Dezhou's low-carbon development status, trends, goals of various sectors (including industry, transport, construction, tourism, agriculture, forestry, renewable energy industry), and sets corresponding development goals and measures.

In terms of carbon management in urban areas, Dezhou has not yet completed its carbon emission inventory. With the support of the LCCC project, it is starting to trace its energy-activity-related carbon emissions and composing an inventory. There are no plans to develop an inventory for non-CO_2 emissions in industrial processes, land use change and forestry, waste disposal and other aspects yet.

Dezhou does not have an integrated long-term plan for renewable energy development except for its solar development plan. No survey or assessment has been done on natural resource endowment. In the promotion of renewable energy utilization, Dezhou focuses on solar energy only, with limited policy/financial support or instructions available for other renewable energy projects such as biomass power generation, CDM project development, etc. Currently local renewable energy production accounts for only a very small proportion in the total energy consumption mix, and there is a need for further support and development.

Dezhou has rather complete data on the gas penetration rate among urban households, the comprehensive utilization rate for industrial solid waste and the sanitary treatment rate for municipal solid waste. The sanitary treatment rate for municipal solid waste reached 98%. Garbage sorting has been achieved at all manual, machinery and fixed collection points. Awareness-raising activities were conducted for waste sorting and reduction. Dezhou has taken the lead in imposing a waste treatment fee based on the amount of waste generated per capita, which induced customers to reduce waste. Dezhou's waste-to-energy incineration project is also progressing well.

In urban development planning, Dezhou has developed plans for roads, traffic, the sewage system, earthquake and disaster prevention, and meteorological disaster prevention, which all take into account ecological and low-carbon development factors. However, no clear low-carbon objectives and action plans have been set in municipal infrastructure planning (e.g. district heating, water-saving, waste water treatment) yet.

China has not yet introduced any green product catalog for green procurement. Dezhou is strictly implementing current national government procurement standards. All government financial support for energy conservation activities is regulated with formal documents which are sent directly to all large enterprises to ensure they are well-informed.

For transparency/accessibility of low-carbon planning and management documents and other related information, Dezhou still needs to further improve its current work. It can increase publicity through the Internet and other forms of media to encourage public participation and to get comments and feedback. Dezhou needs to strengthen low-carbon training for the leadership cadres and the public and enhance the understanding of the low-carbon concept and practice. Dezhou has yet to carry out demonstration of low-carbon projects among communities, schools, hospitals, shopping malls, supermarkets, etc. It is recommended that Dezhou can develop its local evaluation criteria and carry out pilot projects and related appraisals. Such pioneering and experimental work will be beneficial to its low-carbon development.

4.5.2 *Green Economy*

Dezhou is committed to developing its new energy industry. It has developed a non-fossil energy development plan including the Dezhou Rural Biogas Tanks Construction Plan, New Energy Industry Development Plan and other relevant plans. The municipality vigorously promotes the development of the renewable energy industry in order to find new economic growth points. Dezhou has become the world's largest solar thermal industrial base, the national new energy industrial base of the Torch Program, and the national renewable energy demonstration city. According to the Plans, annual sales revenue of the renewable energy/energy-saving industry in Dezhou should exceed 100 billion *yuan* in 2012–2014. Dezhou strives to build the world-renowned China Dezhou Sun City and the Capital City of China New Energy.

Dezhou has strengthened its policy, technological and procurement support for development of the new energy industries, support and consumer support. A special fund has been established to support high-tech enterprises in new energy (energy-saving) industry to help them achieve healthy and rapid development. Dezhou also set up a New Energy (Energy Saving) Industry Promotion Office and the Chinese Sun City Strategy Promotion Committee. It encourages new energy (energy-saving) enterprises to develop research cooperation with talents from larger cities such as Beijing, Tianjin and Jinan so as to enhance their capability for independent innovation. The National Solar Thermal Engineering Technology Center and Solar Thermal Industry Testing Center have already been built in Dezhou. The city conducted more than 20 large national key research projects on solar energy in China's 863 Program as well as 85% of all solar-related National Key Scientific and Technological Research Projects of the Eleventh Five-Year Plan. The municipality actively fosters a positive and healthy ecological culture and advocates green consumption patterns. It vigorously promotes the use of solar energy, ground source heat pumps, new energy vehicles and other new energy products, and has implemented the Solar for One Million Roofs Program, Solar Bathrooms for Rural Area Project, Solar Lighting for Roads and Solar Communities.

By 2011 Dezhou phased out 1,928 sets of obsolete equipment and 15 old production lines. Moreover, 22 small enterprises with serious pollution were closed down, mainly in the fields of cement, thermal power, glass, machinery, chemicals, textiles, clothing and dyeing. Fishes are found again in the four main rivers in Dezhou and people have started to wash their laundry again by the river as water quality improved. The number of days on which urban air quality reaches "good" or "excellent" increased by 5% for five consecutive years. When Dezhou was selected as an exemplary case in Shandong for national inspection in the Haihe River Basin Water Pollution Control Program, it won the two top prizes among seven provinces/cities along Haihe River and among 33 exemplary cities in China's nine major river basins.

Dezhou's government has paid close attention to developing enterprise energy management systems, encouraging enterprises to strengthen their energy conservation management. Based on Dezhou's Key Energy-Using Companies Energy Management System Construction Implementation Plan, the municipality has officially launched the first batch of 16 companies to build energy management systems. Two companies have already installed energy management systems. By the end of 2013 all key energy-using companies with a consumption of over 10,000 tons will establish energy management systems.

There is also room for improvement in Dezhou's green economic development. There is no voluntary emission reduction agreement signed with companies yet; no data is collected to assess the effects of campaigns on reducing chemical fertilizer use and increasing the use of organic fertilizers; biogas utilization rate in rural areas is less than 90% as the amount of livestock manure has been reducing and biogas digesters are not in use; there is not much attention paid to biogas development in livestock farms.

4.5.3 *Green Building*

As early as 2006, Dezhou had already started to implement the national standards of 65% energy saving for residential buildings and 50% energy saving for public buildings. During the Eleventh Five-Year Plan period, Dezhou completed energy-saving retrofits for existing buildings of 1.76 million m², exceeding the 1.5 million m² target. Using the funds from the construction supporting fee and building envelop retrofitting fund, the municipal government provided a subsidy of 45 *yuan* per m² retrofitted. Dezhou was selected as a national pilot city for heat metering reform in 2007 and completed heat meter installation and building retrofit for 1,042,200 m² of existing buildings in 2009, exceeding its target assigned by the Shandong provincial government. In 2010, Dezhou further selected 300,000 m² of existing buildings on six residential blocks for integrated energy saving retrofitting.

Renewable energy application in buildings is a priority in Dezhou. The municipality established the Sun City Strategy Energy Office to be responsible for renewable energy utilization. In 2008 Dezhou commissioned the Chinese Academy of Building Research to prepare the Master Plan on Renewable Energy Application in Buildings (2009–2015), which has set the targets of completing 1.977 million m² with renewable energy application in 2010, and 3.04 million m² in 2011. Dezhou was among the first batch of national renewable energy demonstration cities at the end of 2009 and received a grant of 60 million *yuan*. At the end of the Eleventh Five-Year Plan period, 95% of buildings in the downtown area and 50% of buildings in all urban areas had solar energy applications. In the central urban area of Dezhou, the total solar collector area reached over 600,000 m², saving 100,000 tons of standard coal each year. More than 200 villages in Dezhou have built solar shower rooms.

The green-building-related capacity-building system is rather complete. Dezhou has developed comprehensive green building training materials, covering the design, construction, supervision, construction, quality control, management, etc. The targeted audience is business professionals and government officials.

In fact, some buildings in Dezhou have already reached China's green building standards, such as Himin Group Weipai Tower and Himin Engineering and Technology Institute that are enlisted as national demonstration buildings integrated with photovoltaics, or Tianyuan Tower and other ten residential blocks selected as building energy efficiency demonstration projects by the MOHURD. These projects did not apply for green building certification due to the high cost of application. Dezhou imposes mandatory checks on buildings with solar energy applied. For all buildings with renewable energy applied, their architecture design drawings will be reviewed for approval.

Dezhou is making steady progress on green building management. In its Twelfth Five-Year Plan Dezhou set targets to reduce overall energy consumption per unit building area and to support EMC projects for buildings. According to the Dezhou Green, Ecological, Low-Carbon City Building Twelfth Five-Year Plan,

Dezhou aims in the next five years to have 100 tall buildings installed with solar hot water systems, achieve solar building integration in 100 rural communities for effective improvement of the farmers' living environment, get 100 residential communities installing solar photovoltaic generation system, and build a "green downtown" by implementing high-standard greening projects for the six major parks, ten public green spaces, 38 roads and 100 residential blocks. The Plan further stated that by 2015, the city's energy consumption per unit gross floor area in 2015 should decrease by 20% compared to 2010, district heating for buildings will be all charged according to the actual heat consumed and renewable energy will account for more than 20% of total building energy demand.

Currently energy consumption statistics on large-scale public buildings and government office buildings have been collected according to MOHURD requirements. Dezhou, as a pilot city, has established a rather complete statistical system and has gathered some data on public buildings. But for residential building energy consumption data statistical work still needs to be improved. At present, the central department has not issued its action plan for green building for the Twelfth Five-Year Plan period, while Dezhou has not released its medium- and long-term plans and action programs for green building either. It is recommended that Dezhou should speed up the progress of green building development, establishing pilot projects and increasing the proportion of green buildings.

4.5.4 *Low-Carbon Transport*

Dezhou has introduced its Twelfth Five-Year Transportation Plan which has integrated many low-carbon elements. In addition, Dezhou also developed a Twelfth Five-Year Low-Carbon Transportation Plan. Other specific low-carbon transportation plans and projects are also in the pipeline. Dezhou's existing or upcoming low-carbon transportation-related projects include: renovation of smart toll stations (express toll service), promoting Drop and Pull Transportation and recycling of asphalt pavement residual produced from road reconstruction. The Safety and Technology Division in Dezhou

Transportation Bureau is responsible for energy consumption statistics collection on commercial vehicles (buses, freight, passenger transportation, taxis, etc.). Energy consumption statistics for private cars are not available.

Dezhou is still in the beginning phase in terms of integrated transportation management. Although it has begun public transport information platform construction and planned for bus rapid transit (BRT) projects, no official documents have been formed yet. Dezhou's low-carbon transportation development demand is relatively weak compared to the strong urge for economic development. Public transportation for both freight and passengers is currently sufficient. No systematic evaluation has been carried out on vehicle demands and development trends, nor have strategic plans been developed for pilot projects. It is worth mentioning that Dezhou has established a Transportation Management Division to integrate four tasks previously assigned to different institutions (namely, road maintenance, transportation, traffic and logistics) in accordance with Shandong's provincial requirements.

Dezhou has made outstanding achievements in promoting new energy vehicles in public transit. New energy vehicles in Dezhou are no longer at the pilot project phase but the broad application phase, with the ratio of new energy vehicles in public transport reaching 68.6% and the proportion of new energy vehicles in taxis reaching 95.6%. Currently, Dezhou already has seven natural gas stations, while the Planning Bureau is revising its plan on natural gas and petrol oil stations. Dezhou currently collects data on fuel types used by vehicles through motor vehicle registration. Although the absolute number of new energy vehicles is gradually increasing, training has not yet been carried out for sales staff on knowledge of new energy vehicles.

Solar street lighting system is highly advanced and data on existing street lighting and energy consumption is available. To date, a total of 77.27 km of streets in Dezhou are equipped with an aggregate lighting power of 1,363.74 kW (Law Enforcement Bureau). All road lighting and traffic signals use solar lighting in Dezhou's new districts and Dezhou economic development zones. In the downtown, the current stock of old lamps will be replaced with solar lamps at the end of their life cycle. Based on these data, it can be concluded that renewable energy street lighting has reached a considerably large proportion in Dezhou. In addition, Dezhou initiated other renewable energy or energy-efficient lighting pilot projects, e.g. a company is contracted to install solar lamps and xenon lamps for highways and urban demonstration areas.

The number of low-carbon awareness-raising activities increased and related capacity-building activities were well implemented. Currently, Dezhou carries out seven to eight low-carbon campaigns each year, including Low-Carbon Experience Day on June 14 every year, lectures for government officials in Shandong Province, Car-Free Day, 10,000 People Jogging Together Day and four to five relevant meetings and lectures every year. The scope of people participated is also expanding. Low-carbon transport training is divided into two parts, one on general knowledge for all people in the sector, and the other on professional knowledge and skills for related professionals. Each training activity has a specific theme and easy-to-apply materials. This ensures that all stakeholders have a chance to be trained while professionals trained can quickly apply capacity built during the training in their daily work.

The average commuting time in Dezhou is far shorter than that in first-tier cities. According to the 2008 Road Traffic Plan composed by Dezhou Planning Bureau, Dezhou residents' average commuting time is 22.5 minutes, significantly less than the 50 minutes in Beijing. Travel time for socializing purposes (visiting friends and relatives) is the longest, followed by business travel. From the vehicle perspective, travel time by bus is the longest, followed by company car. Travel times for different purposes show very little difference (see Figures 4.5 and 4.6).

Dezhou residents' satisfaction level with public transport is not very high. According to a survey conducted by the Planning Bureau in 2008, the rate of people who chose "satisfactory" or higher was only 18.5%, while 39.8% of interviewees considered it "so-so" and 41.7% were "dissatisfied or very dissatisfied".

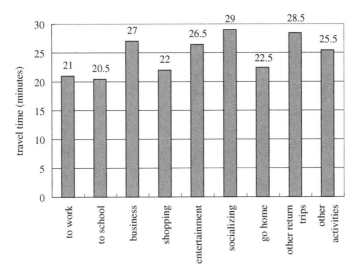

Figure 4.5 Travel Times for Different Purposes in Dezhou.

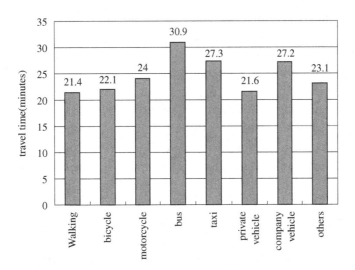

Figure 4.6 Travel Times by Different Ways.

Data source: Dezhou Planning Bureau, 2008 Transportation Plan.

Reasons for dissatisfaction were mainly expensive fares, inconvenient arrival and transfer, and poor punctuality. The number of bus rides per 10,000 residents in Dezhou did not meet the national standard. A survey showed that a quarter of the population considered "vigorously developing public transport" as the most important priority to improve the urban transport infrastructure. Responding to this survey, Dezhou started to pay a high level of attention to public transport services. The municipality carefully considered convenience for arrival and transfer while developing bus station plans, conducted regular assessment on punctuality, convenience and affordability of public transport and took measures for improvement.

4.6 Challenges

From the LCCC Primary Indicator System evaluation results, it can be concluded that Dezhou has made certain achievements in its low-carbon city construction. Its carbon productivity and share of non-commercial renewable energy use and growth trends in absolute terms exceeded the national average. Its economic growth rate is higher than its carbon emission growth rate, indicating carbon emission and economic growth are getting relatively decoupled. A relatively complete low-carbon management system has been established and low-carbon environment indicators are significantly better than average. However, there are still difficulties and challenges.

Dezhou's energy intensity per capita is rather high, and its per capita carbon emission is much higher than the national average. This is because Dezhou is an industrial city. In 2009, the ratio among the three industries in Dezhou was 11.2 : 54.9 : 33.9 — although the proportion of the tertiary industry rose by 1.9%, it is still much lower than the national average of 42.6 and provincial average of 34.1. Looking at Dezhou's industrial structure, we see that the traditional pillar industries (accounting majority share of GDP) are mostly traditional energy-intensive industries with very low willingness to commit to voluntary emission reduction. Although energy management systems are vigorously being promoted, no enterprise has yet to sign any voluntary emissions reduction agreement. The industrial sector accounts for about 85% of total energy consumption in Dezhou. The top ten industries with the highest sales revenue are all energy intensive, including textile, paper and paper products, chemical materials and chemical products manufacturing, non-metallic mineral products, ferrous metal smelting and rolling processing industry, etc., all of which are the top ten energy-consuming industries in Shandong Province. In 2007, Dezhou's energy consumption per unit of industrial added value was 2.17 tce/10,000 *yuan*, 14% higher than the provincial average and 26% higher than the national average. Overall per unit energy consumption decreased gradually, but the reduction was cancelled out by faster growth in output of the high energy-intensive industries such as chemical, power generation, iron and steel, paper making, etc.

In addition, with Dezhou's accelerating urbanization process, the level of consumption and energy consumption is also rising, while low-carbon awareness among residents is still low. Dezhou's number of buses owned per 10,000 residents was only 5.7 standard units in 2010, 60% less than the national average. Most residents prefer non-motorized travel modes. Yet private car ownership is rapidly growing, and there is a strong need to increase investment in public transport and increasing awareness. Dezhou residents' income is also on the rise and there is still a strong impetus to increase household consumption when purchasing power is not fully released. As income levels improve, it may be accompanied by an increase in energy consumption and carbon emissions. The rise in consumption-associated carbon emission will pose challenges to achieving targets of reducing per capita carbon emission.

Based on the assessment of its efforts, it looks like Dezhou also faces enormous challenges. Dezhou's efforts to increase the share of non-fossil fuels in primary energy consumption and to reduce carbon intensity are relatively weak. Scoring higher in these two indicators will require Dezhou to upgrade its energy mix. The general mix usually shows distinct regional characteristics, and is difficult to change only by efforts at the municipal level. Shandong's power grid is coal-dependent where thermal generation accounts for more than 95%. This is the primary reason why Dezhou cannot do very well in these two indicators.

Currently Dezhou is a high-carbon city at a high level of economic and social development, when high economic growth and high energy consumption and carbon emissions coexist. As a relatively underdeveloped city in an economically developed province, Dezhou has the drive to pursue economic growth because of peer pressure, and has inherent desire to recruit business, develop industries and expand its economic scale. It is foreseeable that, for a rather long time to come, the municipality will still desire faster economic development. To achieve a low-carbon transition while maintaining economic growth requires tremendous efforts.

4.7 Conclusion and Recommendations

Dezhou is at the critical phase of transition from a traditional high-energy consumption city to a sustainable low-carbon city, and the five years during the Twelfth Five-Year Plan period is very important. First, Dezhou has significant motivation for economic expansion; second, the municipality proposed the development of a low-carbon eco-city, which is an inherent demand and is in line with the fundamental interests of the people of Dezhou; third, in recent years its development in the renewable energy industry and the Sun City has provided Dezhou a new growth point for the transition from a traditional high-energy consumption city to a sustainable low-carbon city; fourth, such low-carbon efforts also echo China's strategy of scientific development and building a harmonious society.

In the building of a low-carbon city, Dezhou has advantages as well as disadvantages. One highlight is its cluster of renewable energy industries featuring the solar energy industry, whose presence educated governments officials, business professionals and the general public on the concept of low-carbon and sustainable development. Another highlight is its pioneering work and leadership on energy monitoring and management methodologies created during the implementation of energy conservation, which lays a good foundation for the next step in promoting industrial energy saving. Shortcomings include its carbon-intensive fossil-based energy mix, its dependence on electricity supplied by Shandong's coal-generation-dominated grid (99% is coal generation), and its economy heavily relying on the traditional energy-intensive industries. In urban management, carbon management awareness has not been systematically translated into concrete measures and there is a need for integrated overall planning.

(1) Further strengthen government management mechanisms and establish management institutions with clear responsibilities and tasks for low-carbon city development. The Leading Group should be led by city mayors and department heads. Daily work is managed by the Leading Group Secretariat and teams are built for the implementation of energy conservation and low-carbon city construction. The main objective of such an institution is to coordinate the two important goals of economic development and energy saving. Take planning as the starting point — integrate low-carbon concept and measures into the planning of all sectors, and provide effective tools to monitor energy consumption and carbon emissions for the implementation of the plans. Conduct annual reviews in order to avoid and to correct mistakes. To reduce the pressure of reaching energy conservation targets, efforts need to be paid to each phase of the cycle.

(2) Consolidate the further development of existing energy management systems. Dezhou is the pioneer in energy management systems and achieved initial results, which is a good start for low-carbon city administration. During the Twelfth Five-Year Plan period, Dezhou plans to establish a comprehensive energy management system for all major energy-using industrial companies to assist them with energy management and energy efficiency improvement with this long-term mechanism. The next step is to develop an energy management system suitable for businesses and to promote it among construction, transportation and other public service institutions, so as to become a provincial leader and even a leader in the country.

(3) Compile a low-carbon inventory and develop low-carbon city management tools as the basis for policy-making. According to

the LCCC Indicator System recommendations, the tool should include three databases or software: first, urban carbon emission inventory which is a top-down model established with statistics collected and used for monitoring and forecasting the trend of different sectors; second, energy consumption data on different types of buildings; third, an emission inventory for different modes of transportation, including buses, private cars, taxis and so on. Take transportation for example. Dezhou has linked vehicle exhaust testing with vehicle annual checks to strengthen regulation on tailpipe emissions and encourage the use of energy-efficient new energy vehicles.

(4) Increase public awareness campaigns and propaganda about a low-carbon city to enhance the city's profile. Strengthen the brand of China Sun City — Low-Carbon Dezhou and extend the connotation of China Sun City. Gaining a good reputation internationally will help Dezhou to get better opportunities and conditions for its low-carbon development. Consumption behaviors featuring low-carbon activities is another important sign for successful low-carbon city development. Dezhou residents' current per capita consumption level is still low in Shandong Province and its purchasing power is still to increase, which means this is the easy phase to change consumer behavior patterns. Involve the broader public to raise low-carbon awareness and advocate for ecology, energy saving, clean consumption patterns as a fashion — this can help gradually raise the living standards of residents while keeping energy consumption and carbon emissions stable. Take advantage of Dezhou's favorable geographical position and make the city a platform for harmonious coexistence between human and nature, favorable conditions to establish a three hours traffic circle within the city, a place that attracts business and capital with a preference for ecological and environmental advantages.

(5) Strengthen international cooperation to get external funds and technical guidance and support. Dezhou is a developing city with prominent conflicts between economic development and carbon emissions. It lacks the necessary capacity to develop a low-carbon city at present and needs to seek all levels of technology and financial support

from the domestic government and the international community for its low-carbon projects. In addition, cooperation projects usually bring in the advanced knowledge and successful experience of other regions. Developed European countries have made remarkable achievements in their low-carbon city construction and have accumulated a wealth of low-carbon city construction technology, methodologies and experience. Dezhou can pursue further international cooperation based on the LCCC project. China's first batch of low-carbon pilot regions — the five provinces and eight cities — has been carrying out their exploration work in accordance with the NDRC's requirements. Dezhou needs to further learn from both domestic and international experience.

References

1. Institute for Environmental Policy and Environmental Planning, China People's University. 2010. *Twelfth Five-Year Low-Carbon Development Plan for Dezhou.*
2. Cui Yuqing. 2010. *Dezhou Low-Carbon Development Assessment.* Research report of the Institute for Urban and Environmental Studies, CASS.
3. Dezhou CPC Committee, Dezhou People's Government. 2010. *Guidelines on Accelerating Low-Carbon Economy Development* (#Defa [2010]12).
4. Dezhou People's Government. 2010. *Report on Energy Conservation Work in 2009.*
5. Dezhou Commission of Industries and Information. 2010. *Report on Biotech and New Energy Industry Base Construction.*
6. Dezhou Commission of Industries and Information. 2010. *Report on New Energy Industry Development.*
7. Yang Kai. 2011. Public building energy consumption in heating area of Northern China. *Wall Innovation And Building Energy Efficiency,* 3: 57–58.
8. China Academy of Building Research, Dezhou Housing and Urban-Rural Construction Bureau. 2010. *China Sun City-Dezhou Construction Standards Research Reports* (discussion paper).
9. Dezhou Planning Bureau. 2008. *Dezhou Comprehensive Transportation Plan* (2008-2020).
10. Institute for Urban and Environmental Studies in CASS, China Low-Carbon City (LCCC) Project Office. 2012. *China Low-Carbon Indicator System — Methodology Report.*

Appendix: Recommended Low-Carbon Action Plan for Dezhou

#	Recommended Action	Details	Responsible Dept.	Priority	Deadline (years)
I.	*Urban management*				
1.	Preparation of comprehensive carbon emissions inventory of the city	Establish emission monitoring, statistical, feedback and long-term regulatory mechanisms. Compile emissions inventory based on statistics system	LCMO leading, DRC and SB supporting	High	2
2.	Develop emissions inventory preparation programs and capacity-building programs	Responsible departments issue specific program, and invite comments from experts	LCMO leading, SB supporting	High	0.5
3.	Establish evaluation mechanisms for the use of funds allocated	Establish feedback channels and periodic verification of funds, compare with expected results	FB leading, other institutions supporting	Low	0.5
4.	Integrate green procurement into government planning	Develop Green Product Directory based on green procurement list to regulate procurement	DRC	High	1

(Continued)

(Continued)

#	Recommended Action	Details	Responsible Dept.	Priority	Deadline (years)
5.	Prepare green products directory and supporting files	Prepare the directory based on principles such as energy/water saving, low pollution, low toxicity, renewable and recyclable with reference to the country's list	Government Purchase Office of Dezhou	High	1
6.	Establish information platform for transparency of low-carbon planning and management information	Establish information platform via the Internet, radio, newspapers, etc., to publicize low-carbon planning and management information	LCMO	High	0.5
7.	Collect and publicize low-carbon-related information including policies, initiatives, projects, activities, etc.	Collect and publicize low-carbon-related information including policies, initiatives, projects, activities, etc.	LCMO	High	1
8.	Build channels for feedback on low-carbon planning and management	Through newspapers, administrative agencies and other communication channels	LCMO	Medium	0.5

(Continued)

Continued

#	Recommended Action	Details	Responsible Dept.	Priority	Deadline (years)
9.	Collect comprehensive data on renewable energy production	Collect comprehensive data on renewable energy production	LCMO leading, HURD, DRC, SB supporting	High	1
10.	Set clear water-associated energy management targets	Set feasible targets through surveys and discussions	Municipal utilities administration	Medium	0.5
11.	Set clear regulations for sewage treatment plant energy management	Responsible department issues explicit regulation guidelines in accordance with local actualities	HURD — waste water division	High	1
12.	Set clear waste-water-related energy management targets	Set feasible targets and integrate them into urban planning	HURD — waste water division	High	1
13.	Develop water conservation action plan	Responsible departments set targets and action plans respectively for residential and industrial water use reduction	Residential: Public Utilities Bureau; Industrial: IIC	Medium	1

(Continued)

(Continued)

#	Recommended Action	Details	Responsible Dept.	Priority	Deadline (years)
II.	**Green economy**				
14.	Get GHG emission statistics for industrial processes	Establish mechanisms for emission monitoring, statistics collection, feedback and long-term regulation of industrial enterprises. Build emissions inventory accordingly	EPB leading, IIC, SB, DRC supporting	High	2
15.	Develop systematic DSM planning	Develop systematic DSM planning from consumption perspective	IIC — Energy Division	High	1
16.	Promptly publicize energy monitoring results	Promptly publicize energy monitoring results	LCMO — Energy Monitoring Committee	Medium	0.5
17.	Set specific targets for renewable energy use among industries	Set specific targets for renewable energy use among industries based on urban master plan	DRC, IIC — Energy Saving Office	High	1
18.	Regularly write and publish reports of industrial solid waste utilization	Set special projects and funds to write and publish reports of industrial solid waste utilization	IIC — Energy Saving Office	Medium	1

(Continued)

(Continued)

#	Recommended Action	Details	Responsible Dept.	Priority	Deadline (years)
19.	Appoint full-time energy manager with clear responsibilities in public institutions	Leading institution should ensure that at each responsible department, a full-time energy manager with clear responsibilities is appointed	Government Service Bureau, Business Bureau, HURD	High	1
20.	Sign voluntary emissions reduction agreements with service companies to promote low-carbon consumption	Develop template, conduct publicity campaign and encourage service companies to sign voluntary emissions reduction agreement	Business Bureau, Tourism Bureau	High	2
21.	Provide policy incentives to encourage energy service companies to register and operate	Provide (both material and non-material) incentives to encourage energy service companies to register and operate	DRC, IIC	Medium	1
22.	Provide relevant information and technical support for producers regarding agriculture and forestry residue treatment and reuse	Provide relevant information (including manuals, notices and documents) and technical support for producers regarding agriculture and forestry residue treatment and reuse	AB – Eco & Environment Protection Division, Forestry Bureau	High	2

(Continued)

(*Continued*)

#	Recommended Action	Details	Responsible Dept.	Priority	Deadline (years)
23.	Issue policies that encourage farmers to get certified for green/organic/pollution-free products	Issue policies that encourage farmers to get certified for green/organic/pollution-free products	AB	High	1
24.	Develop a plan to achieve agricultural energy savings	Responsible department develops a plan with specific measures	AB — Rural Energy Office	High	2
III. Green building					
25.	Develop a sound work plan to collect building energy statistics (including the statistical classification for all building types)	Establish mechanisms for emission monitoring, statistical, feedback and long-term regulation. Compile emissions inventory based on statistics	HURD — BERO	High	3
26.	Set clear targets for heat metering reform and existing building energy retrofit according to the actual situation	Set clear targets, including both short-term and long-term ones	HURD — retrofit office, and Municipal Infrastructure Bureau — Heat Reform Office	High	2

(*Continued*)

(Continued)

#	Recommended Action	Details	Responsible Dept.	Priority	Deadline (years)
27.	Prepare energy use plan and action guidelines for new buildings in new urban area	Prepare energy use plan and action guidelines for new buildings in new urban area	HURD	High	1
28.	Preparation of new urban planning, new buildings, energy use, planning, action guidelines	Further promote building integration retrofit for solar and other renewable energy. Organize campaigns and provide policy and financial incentives	HURD, Solar City	High	3
29.	Integrate renewable energy applications into administrative regulatory review process (except drawing inspection phase)	Integrate renewable energy applications into administrative regulatory review process	HURD — BERO	High	2
30.	Prepare green building capacity-building plan	Prepare green building capacity-building plan from the government's perspective	HURD — BEEST	High	1
31.	Conduct building energy assessment after the completion of the building for acceptance	Responsible department issues regulations or policies to enforce energy assessment before acceptance	HURD — BEEST	High	3

(Continued)

(*Continued*)

#	Recommended Action	Details	Responsible Dept.	Priority	Deadline (years)
32.	Develop green building marketing plan preparation	Develop green building marketing plan preparation	HURD — BEEST	High	2
IV.	**Low-carbon transport**				
33.	Start preparing carbon emissions inventory for the transport sector	Establish emission monitoring, statistical, feedback and long-term regulatory mechanisms. Compile emissions inventory based on statistics system	TB — TMDSTD	High	2
34.	Clarify funding sources and the responsible departments to implement integrated transport	Clarify funding sources and the responsible departments to implement integrated transport	TB — TMDSTD	High	0.5
35.	Train procurement staff on new energy vehicles	Conduct regular training activities	TB — TMDSTD	Medium	2
36.	Collect data on the number of existing road lamps and their energy consumption	Collect data on the number of existing road lamps and their energy consumption	Law Enforcement Bureau — Road Lamp Office	High	1

(*Continued*)

(*Continued*)

#	Recommended Action	Details	Responsible Dept.	Priority	Deadline (years)
37.	Formulate guidelines and action plan that gives priority to public transport	Responsible departments take the lead to formulate guidelines and action plan that gives priority to public transport	TB, bus companies	Medium	1
38.	Set specific measures to increase public transport service, including punctuality, convenience, comfort level and affordability	Conduct surveys and expert seminars to develop such specific measures	TB, bus companies	High	1
39.	Develop integrated plan for slow traffic key infrastructures	Develop integrated plan for slow traffic key infrastructures	Planning Bureau	High	1

Abbreviations

DRC: Development and Reform Commission
HURD: Bureau of Housing, Urban and Rural Development
IIC: Industrial and Information Commission
FB: Financial Bureau
PB: Planning Bureau
AB: Agriculture Bureau
TB: Transport Bureau
BERO: Building Envelop Retrofit Office
BEEST: Building Energy Efficiency Science and Technology Department Energy Monitoring Committee, Public Building Energy Management Agency
TMDSTD: Traffic Management Division Safety and Technology Department

Chapter Five

Assessment of Kunming's Low-Carbon Development

5.1 Facts of Kunming

Kunming is located in the middle of the Yunnan-Guizhou High Plateau in south-western China, with a longitude of 102°10' to 103°40' and a latitude of 24°23' to 26°22'. It is surrounded by mountains with Dianchi Lake at its south. The altitude of the urban downtown is about 1,891 m above sea level. Overall the terrain shows tiered elevation from the higher northern area to the lower south. Meanwhile it is higher in the middle and lower on the east and west sides. Kunming's jurisdiction covers six districts (Panlong, Wuhua, Xishan, Guandu, Chenggong and Dongchuan) and four counties (Fumin, Songming, Yiliang and Jinning), three autonomous counties (Xundian County, Shilin County and Luquan County), a city (Anning) and three national development zones (High New Technology Development Zone, Economic and Technological Development Zone, Dianchi Tourist Resort) (Figure 5.1). It has a total area of 21,012.54 km² and a population of 6.4392 million (2010).

Kunming has a subtropical low latitude, high plateau/mountain monsoon climate. With warm air coming from the Indian Ocean in the southwest, it has long sunshine hours and a short frosty period. In 2010, on average Kunming had a temperature of 16.4 °C, 2,136.8 hours of sunshine, 869.1 mm of rainfall, humidity of 69% and 30 frost days. It is the famous Spring City and Flower City in China because of its mild climate and minimum temperate difference throughout the year.

Figure 5.1   Kunming's Jurisdiction.

From the geographical point of view, Kunming is the gateway to countries in Southeast Asia, South Asia, the Middle East, Southern Europe and Africa, reaching the east coast through Guizhou and Guangxi, connecting with central China through Sichuan and Chongqing, close to Vietnam, Laos, Thailand and Cambodia to the south, and in close proximity to Myanmar, India and Pakistan to the west. With this unique geographical advantage, it has become the ideal place for international trade. The two trade centers in Kunming — ASEAN Free Trade Area (10 + 1) and the Pan-Pearl River economic cooperation zone (9 + 2) — make the city a platform connecting international and domestic economies. Kunming already has several major railway lines including the Guiyang-Kunming railway, Chengdu-Kunming railway, Nanning-Kunming railway, Kunming-Dali railway and Yunnan-Vietnam Railway. The Mekong waterway is now available for water transport, connecting China to five ASEAN countries. It is now a major channel for importing and exporting.

Kunming's six pillar industries are tobacco, chemical, pharmaceutical, metallurgy, machinery and power generation. There are highly competitive enterprises with local characteristics in tobacco processing, machinery manufacturing, biotechnology, agriculture flowers and other industries.

5.2 Background

5.2.1 *Implementing the Country's Low-Carbon Strategy*

Cities are the major carrier of the formation and development of modern civilizations, where the population and industries gather. Meanwhile they are also the center of carbon emission. The Chinese government has set an explicit goal in the Twelfth Five-Year Plan of "green development and building a resource-saving and environment-friendly society", and that "in the face of increasing resource and environmental constraints, we must be aware of environmental crisis, and establish green, low-carbon development philosophy, focus on energy conservation, enhance incentives and regulations, accelerate the construction of resource-saving and environment-friendly production methods and consumption patterns, build capacity for sustainable development, and improve the level of ecological civilization". Faced with the dual challenges of economic transformation and "green revolution" domestically and globally, China needs to seize the international trend of a low-carbon economy and green development and turn pressure into opportunities.

As one of the developing countries actively responding to climate change, China has initiated various programs for its low-carbon city development. At the national level, China identified five provinces and eight cities as low-carbon city pilot regions, where low-carbon development planning and supporting policies are developed, and low-carbon industries such as new energy and other strategically important industries are being fostered to reduce emissions and improve environmental protection. At the local level, cities are the main carrier of national economic development, and national low-carbon economic goals are ultimately allocated and to be achieved at the city level. Therefore low-carbon development is an important national initiative for addressing climate change, but also an important future direction for cities' development and transformation.

5.2.2 Rich Resources for Low-Carbon Development

First, Kunming has a beautiful environment with high vegetation coverage. It is an important region for China's natural forest protection projects. Kunming's unique geographical location and rich forest resources make it an extensive carbon sink. In 2008 Kunming had a total forest area of 1,365.54 *mu*, accounting for 43.33% of the total area of the city. Forest coverage rate was 45.05%. The city has many lakes, rivers and large urban green spaces. These resources lay the foundation for developing tourism and low-carbon agriculture.

Second, Kunming has abundant hydropower, solar and other renewable energy resources. Kunming's annual solar energy rate is 80%, with 2,250 sunshine hours on average and solar radiation of $5,461 \, \text{MJ/m}^2$. Solar energy in Kunming features long hours, high intensity and even distribution, and as a result, there is broad application of solar devices. Penetration of solar water heaters achieved 40%. In addition, there are also wind energy and water resources available. Kunming's mild climate implies lower energy demand and easier low-carbon conditions. In 2010 non-fossil fuel use accounted for 24.0% in total primary energy consumption, while the share of renewable energy in total primary energy consumption reached 18.09%, much higher than the national average.

Third, Kunming has a temperate climate. It has no severe winters or sweltering summers. This means the energy demand for heating and cooling is very small. Kunming Tourism Bureau has advertised itself to people across China as follows: "Come to Kunming for the winter where no air conditioning is needed".

5.2.3 *Seizing the Early Opportunity for Low-Carbon Transition*

As the capital city of Yunnan Province, Kunming is the leading city for economic development in the province, contributing about one-third of provincial GDP. Meanwhile Kunming is the largest regional energy consumer, accounting for about 25% of Yunnan's total energy consumption. Therefore, Kunming's low-carbon city construction undoubtedly plays a critical role in the province's decarbonization efforts.

Kunming is located in the southwest of China and its level of economic development still lags behind that of eastern coastal cities. Energy-intensive industries, including iron and steel, non-ferrous metal production, chemicals, building materials, etc., are still accounting for a large share in its industrial structure. The energy mix of Kunming is coal-dominated, with a low proportion of oil, gas and clean energy. Efforts to develop new energy or energy-saving technology and products are also insufficient. As a result, for its low-carbon city development, Kunming can try to upgrade its energy mix, increasing the share of hydropower and natural gas while decreasing the share of coal. It should also aim to achieve the transformation from traditional industries to a low-carbon one, cultivating new low-carbon industries and conducting pilot projects.

Kunming enjoys a favorable environment for low-carbon city development at present. First, the central government attaches much attention to climate change and offers significant strategic opportunities for low-carbon development. Second, Yunnan Province was listed as one of the first batch of low-carbon pilot provinces, while Kunming also became one of the second batch of low-carbon pilot cities in 2012 — this will help it to get support for national policies to promote the development of the province's low-carbon economy. Third, with the new round of national strategies including "development in the west", "eastern industrial transfer to west" and Yunnan's "building strong economy and strong culture in Yunnan" strategy and "building central-Yunnan economic zone" strategy, Kunming faces various opportunities for its economic transition.

5.3 Practice and Highlights

Kunming is located in an important geological position and has abundant solar energy and other renewable energy resources, unique landscapes, a beautiful environment, a pleasant climate and a booming tourism industry. These lay good foundations for its low-carbon city development. Kunming has been implementing a variety of programs to transition to low-carbon development, and has achieved considerable success and experience.

5.3.1 *The Municipality Has Strong Political Will for Low-Carbon Development and Pays High Attention to the Leading Role of Planning*

Kunming municipal government established the Energy Conservation Office, Low-Carbon Construction Leading Group and Low-Carbon Construction Document Preparation Work Leading Group, and a few other institutions headed by the Deputy Mayor of Kunming such as the one in the Agricultural and Business Recruitment Bureau. Kunming issued the Document on the Construction of Low-Carbon Kunming in 2010, which set the explicit goal of reducing unit GDP carbon dioxide emissions by 40% in 2020 relative to 2005, and illustrated the strategy of pursuing low-carbon development through building a low-carbon community, low-carbon transport system and low-carbon society.

During the Eleventh Five-Year Plan period, Kunming developed a number of policies to support low-carbon development, including Opinions of Kunming CPC Party Committee and People's Government on Accelerating Development and Utilization of Renewable Energies with Special Focus on Solar Energy and Biomass, Opinions of Kunming People's Government on Strengthening Energy Conservation Implementation, Kunming's Comprehensive Work Scheme for Energy Saving and Emission Reduction, Opinions of Kunming People's Government on Further Strengthening the Implementation of Energy Saving and Emission Reduction, and Kunming Energy Conservation Target Assessment and Rewarding/Sanction Measures (Interim).

Kunming has also introduced the Kunming Low-Carbon Economy Development Master Plan (2011–2020), setting the clear targets of achieving by 2015 a 35% reduction in per unit GDP carbon dioxide emissions from the 2005 level, and a 20% reduction from the 2010 level, and a 45% reduction in carbon dioxide emissions per unit of GDP from the 2005 level by 2020. The plan also proposed a series of supporting policies including industrial policy, price policy and fiscal policy. This document helped promote all aspects of low-carbon work and set the city to embark on a systematic, regulated and healthy track for future development.

In order to be enlisted in the country's second batch of low-carbon pilot cities, Kunming released the Implementation Scheme for Kunming's Low-Carbon Development on the basis of the previously composed *Low-Carbon Kunming Survey Report*. The Scheme divided Kunming's low-carbon development into two phases: the comprehensive construction phase (2011–2015) and the consolidation and enhancing phase (2016–2020). It clarified major tasks and assigned them to specific departments and specific officials. In order to further smooth its low-carbon city development, the Kunming Low-Carbon Development Research Center further prepared the Initial Implementation Plan for the Low-Carbon Pilot Program. After two rounds of national assessment, Kunming successfully became enlisted in the second batch of low-carbon pilot cities in November 2012.

In addition, Kunming has conducted a lot of basic research in GHG emissions, emission inventories preparation and carbon emission reduction potential analysis. The government issued the Kunming Waste Treatment Associated Greenhouse Gas Emissions Inventory (2005–2010), Kunming Carbon Mitigation Potential And Proposed Measures, Low-Carbon Kunming Survey Report, Kunming Low-Carbon Transportation Planning and Kunming Low-Carbon Economy Development Master Plan as the output of 18 research projects.

5.3.2 *Promoting Application of Renewable Energy in Various Fields*

Kunming plans to adjust its industrial structure and build a solar industry that is leading in Yunnan and a showcase in the country, to increase the share of hydropower, oil and gas in the city's energy mix, to reduce the share of coal consumption, and to reduce carbon intensity in smelting, chemicals, building materials, energy and other key industries. It is planned that by 2015, the production capacity will reach 3,000 tons/year for solar silicon material, 500 MW/year or 1,000 MW in total for solar cells, and moreover, solar water heaters will be installed in 1,200 m² of building space and at least six wind farms will be built with a total installed capacity of more than 160 MW.

Kunming has formulated a local building energy efficiency evaluation and certification system and prepared the Kunming Building Energy Efficiency Plan for the Twelfth Five-Year Plan Period and Decisions on Promoting Application and Management of Solar Water Heaters (#Kunrenfa(2009)95). This policy requires all new buildings to integrate solar water heaters at the design and construction phase, and aims to achieve full penetration at all buildings in the urban downtown. It also sets the target of having 95% of buildings in the newly constructed area integrated with solar water heaters, while 70% of buildings in the total urban area and 35% of buildings in the rural area integrated with solar water heaters by 2015. It also sets the target of more than 200 MW or 100,000 m² of solar photovoltaic applications in Kunming and the ratio of energy-saving building materials used in all building materials for new construction to reach 50% or more by 2015.

Kunming is promoting new energy vehicles in public transportation, including new energy buses, sanitation trucks, postal vehicles, etc. In early 2009 Kunming was listed as one of the 13 energy-saving and new energy vehicle demonstration pilot cities, the only capital city in western China. On July 1, 2010, 60 hybrid buses were rolled out in Kunming for the no. 90 and no. 66 bus lines, becoming the province's first batch of new energy bus in operation. Kunming plans to have 1,000 hybrid vehicles in public buses, taxis, government cars, sanitation trucks and postal vehicles by 2012 and earmarked 131.2 million *yuan* to support new energy vehicle application by 2015. In 2011, Kunming's total energy consumption was 21,786,800 tons equivalent of coal, while its renewable energy use amounted to 3,941,500 tons. The next year, Kunming Transport Bureau further initiated eight projects, including the natural gas pilot project, the vertical green parking garage construction, the highway energy-saving project, the public bike project and the green shipping construction project.

5.3.3 Develop Low-Carbon Industries and Upgrade Industrial Structure Based on Local Advantages

With its rich solar and biodiversity resources, pleasant climate and good air quality, Kunming is in a unique position and has a very good foundation for developing its biomass energy, low-carbon agricultural and eco-tourism and service industries.

The ratio among the three industries in Kunming went from 7.2 : 43.8 : 49 in 2005 gradually to 5.7 : 45.3 : 49 in 2010. Its industrial structure has been upgrading, with the total output of high-tech industries increasing more than 20% annually. Because of its continuous efforts in energy conservation and development of new energy industries, energy-saving technologies are widely applied in steel, chemicals, nonferrous metals and other energy-intensive industries. Solar generation, wind generation and waste-to-energy generation technologies are also widely applied in various industries including energy, construction, municipal infrastructures, etc.

For its low-carbon city development, Kunming set a strategic plan of building a full ecological agriculture industry chain based on its own strengths. The government actively promotes green and organic agricultural practices, and tries to reduce chemical fertilizer use by offering technical training, disseminating fertilization soil testing cards, introducing environmentally friendly fertilizers and advocating for comprehensive scientific farm management measures. Meanwhile, Kunming is vigorously developing the flower industry, which is low-energy, low-emission and low-pollution. Kunming is one of the best areas suitable for growing flowers in the world, and its flower exports account for 58.3% of all exports in Yunnan Province. In the 2010 Dounan Flower Market and the Kunming International Flower Auction Trading, the daily trading volume of fresh-cut flowers was over 11.10 million, an increase of 22.8% compared to the end of the Eleventh Five-Year Plan period and accounting for nearly 80% of the total fresh-cut flower trade volume in Yunnan Province, and 38.7% in the country. Kunming has become the country's largest production base and the most influential distribution center for fresh-cut flowers.

The tourism industry is in line with the low-carbon lifestyle and can easily integrate low-carbon technologies. It has inevitably become a favorite for low-carbon economy development. Tourism is one of the pillar industries in Kunming. In November 2011, the Kunming Tourism Bureau launched the Green Kunming, 100 Days Low-Carbon Campaign, and initiated a series of large events such as Meet in the Spring City, Feel the Charm of Kunming to promote low-carbon tourism in cooperation with several website companies.

In February 2012, Kunming was enlisted in the second batch of low-carbon transport system construction pilot cities. The municipality proposed three development strategies: to develop a bus-based urban transport system, to develop intensive urban road network strategies and to support slow traffic development. Kunming became the country's first city to apply the bus rapid transit (BRT) system in the early 1990s. Now, with the BRT system in operation, the number of passengers riding public transport vehicles has significantly increased. Kunming plans to build 22 public bike points in Chenggong New Area and plans to invest in 1,000 bikes by 2012 and 60,000 in total over the next six years.

5.3.4 *Environmental Protection*

Carbon emission is not the only concern of Kunming — improvement of the environment is also seen as an essential part of sustainable development. In order to improve the water quality at Dianchi Lake, Kunming planted water hyacinths covering $26 \, km^2$, which are expected to absorb water and air pollutants while improving soil quality. The city also further strengthened its ban on fishing in the lake and cancelled the original two-month period for fishing in a year. In its "one lake and two rivers" watershed, a ban on raising livestock has been fully implemented to further reduce agricultural nonpoint source pollution. Kunming is also one of the pioneer cities in China to develop technical specifications for urban rainwater collection and utilization. With the implementation of such policies and measures, the number of water-saving enterprises (institutions) reached 17.6% in the total, the industrial water recycling rate went up to 85% and the water quality of surrounding rivers improved significantly. Ecological conditions in Dianchi Lake also improved, with notable watershed environmental benefits and ecological benefits significantly increased.

In addition to reducing carbon emissions, it is also necessary to increase carbon sinks. Kunming City has maintained large green areas, with particular emphasis on ecological isolation belt construction — eight ecological buffer zones were constructed in the surrounding mountains, both working as an isolation belt between different urban clusters and helping to improve the urban environment and increase carbon sinks. Moreover, Kunming built wetlands and forests along the 35 rivers flowing into Dianchi Lake to increase the number of carbon sinks and to reduce pollution in the lake. Kunming also implemented projects to upgrade low-efficiency forests. It carried out a series of ecological improvement projects, including vegetation restoration for the five mining areas, demolishing illegally built buildings and conducting greening projects near Dianchi at the mountain area, building urban ecological green belts/green channels/water-retaining forest/green villages and implementing desertification control. By the Eleventh Five-Year Plan period, the city completed a total of 2.789 million *mu* of afforestation, 1.585 million *mu* of public-interest forests for the Natural Forest Protection Program, banned exploitation of 2.5119 million *mu* of mountain forest, converted 0.748 million *mu* of farmland back to forest and constructed 20,919 *mu* of lakeside forest, 19,080 *mu* of lakeside wetlands, 11,220 *mu* of inner-lake wetlands and 3,086 *mu* of river mouth wetland. The total forest area of Kunming reached 904,600 hectares and forest coverage rate reached 45.05%. The forest stumpage volume reached 39,413,600 m³, with a rate of 36.83% of urban green space, green coverage rate of 41.34%, and per capita public green space reached 11.5 m².

5.4 Status Quo and Efforts

Kunming is one of the pilot cities in the Sino-Swiss joint project, the Low-Carbon City in China project (LCCC). Based on the LCCC Primary Indicator System and related logical framework, the report evaluates Kunming's low-carbon development from two perspectives — status and level of effort. Major data sources include: (1) a variety of publicly available statistics, government documents, statistical bulletins; (2) field research and interviews with relevant departments and stakeholders; (3) domestic and international relevant low-carbon city index system research results.

The LCCC Primary Indicator System includes 15 indicators in 5 categories: economy, energy, infrastructure, environment and society. Among them, the economic indicators reflect a city's low-carbon economic development stage and implementation of the country's overall energy reduction targets; low-carbon energy indicators reflect regional low-carbon energy resources as well as the city's effort in upgrading its energy mix; infrastructure indicators show the urban infrastructure level and low-carbon consumption levels. Environmental indicators assess a city's green development. Indicators on society evaluate consumption patterns and social equity. Specific evaluation results are shown in Table 5.1.

5.4.1 *Low-Carbon Economy*

In accordance with evaluation by the LCCC Primary Indicator System, Kunming's carbon productivity in 2010 was 0.377 in terms of 10,000 *yuan* GDP/tCO_2, an increase of 32.7% over 2005. Compared with the national average of 0.41, it reached 92% of the national average level. This shows that carbon productivity in Kunming is rather low, and has large room for improvement.

During 2005–2010, energy consumption in Kunming increased year on year, with energy consumption per unit of GDP declining by 23.52%, a higher rate than its original target of 17.6%. This shows its energy consumption was declining relative to its economic development, indicating an improvement. However, in terms of absolute energy consumption per unit of GDP, Kunming's figure was 1.128 tce/10,000 *yuan* GDP in 2010, still 9.2% higher than the national average. Reducing energy consumption per unit of GDP and upgrading its industrial structure is critical for Kunming's low-carbon development.

During 2005–2010, the country's carbon decoupling index was 0.86, while it was 0.732 in Kunming, both showing a relative decoupling trend.

Table 5.1 Low-Carbon Development Assessment of Kunming.

Nos.	Indicators	Unit	Kunming		China		Effort
			2005	2010	2005	2010	
	Economy						
(1)	Carbon productivity	10,000 *yuan*/ton CO$_2$	0.284	0.377	0.325	0.413	97%
(2)	Energy intensity	tce/10,000 *yuan*	1.475	1.128	1.276	1.033	100%
(3)	Decoupling index	—	0.732		0.85		80%
	Energy						
(4)	Non-fossil energy in primary energy consumption	%	15	18.09	6.8	8.6	68%
(5)	Per capita non-commercial renewable energy use	kgce/per capita	78	109	16.5	26.7	100%
(6)	Carbon intensity of energy	ton CO$_2$/tce	1.973	1.887	2.405	2.341	75%
	Infrastructure						
(7)	Energy consumption per unit building area for public buildings						
	Government	kgce/m^2	8.6	13.90			
	Large size	kgce/m^2	11.03	18.64	28.02	23.86	50%
	Medium/small	Standard units	7.31	5.132			
(8)	Energy consumption per building area for residential buildings	kgce/m^2	%	4.53	29.3	27.4	50%
(9)	Ratio of green transport	Standard units	%	12.9	8.6	9.7	100%

(*Continued*)

Table 5.1 (*Continued*)

Nos.	Indicators	Unit	Dezhou		China		Effort
			2005	2010	2005	2010	
	Environment						
(10)	Percentage of days with API less than 100	%	kg	100	51.9	81.7	80%
(11)	Forest coverage rate	%	%	113.8	149.8	105.04	94%
(12)	Domestic water consumption per capita per day	kg	ton CO_2/ per capita	45.05	18.21	20.36	100%
	Society						
(13)	Urban-rural income ratio	%	—	3.25	3.22	3.22	100%
(14)	Per capita CO_2 emission	ton CO_2/per capita	kgce/m^2	8.13	4.34	5.67	50%
(15)	Low-carbon management institution	—	None	Rather good	None	Rather good	100%

Note: Indicator (5) mainly refers to biogas and solar energy; indicators (7) and (8) are data from 2009 and 2011 respectively.

5.4.2 *Low-Carbon Energy*

Located in southwest China and rich in hydropower resources, Kunming saw an increase in the ratio of non-fossil fuels in primary energy consumption during 2005–2010, rising from 15% up to 18.09%, much higher than the national average of 8.6% in 2010.

In recent years, Kunming actively promoted solar energy, biomass and municipal waste generation. As a result, per capita non-commercial renewable energy usage is relatively high.

During 2005–2010, Kunming's carbon intensity of energy dropped from 1.973 tCO_2/tce down to 1.887 tCO_2/tce, lower than the national average. This has to do with its hydro-rich energy mix and relatively high energy efficiency.

5.4.3 *Low-Carbon Infrastructure*

Most parts of Yunnan Province belong to the "fifth climate zone" of China featuring a mild climate. Kunming also has a humid and temperate subtropical climate with the following characteristics: (1) big temperature difference between day and night; (2) modest fluctuations in temperature around the year with the average temperature around 15 °C. The energy consumption per unit area of both public buildings and residential buildings in Kunming was notably lower than the national average. According to analysis based on relevant climate indicators, Kunming's regional architectural design strategies need to consider winter heating but not summer cooling. Kunming is rich in solar energy resources, so in winter heating design it should take full advantage of this green energy. The use of passive solar design is not yet very economically viable for broad application; however, other technologies can be considered such as increasing wall insulation.

During 2005–2010, the number of buses owned per 10,000 people increased from 8.2 to 12.9 standard units, while nationwide in the same period it increased from 8.6 up to 9.7 standard units. This is higher than the national requirement of 10–12.5 standard units/10,000 people given by the Ministry of Construction in the Urban Traffic Planning and Design Codes.

5.4.4 *Low-Carbon Environment*

In 2005–2010, the percentage of days with an API lower than 100 rose in Kunming, from 99.45% to 100%. During the same period, the percentage increased from 51.9% to 81.7% nationwide. In this sense Kunming performed better than the national average.

In 2005–2010, the national per capita domestic water consumption decreased from 149.8 kg/person in 2005 down to 104.04 kg/person in 2010. In Kunming it decreased from 124.1 kg/person in 2005 down to 113.8 kg/person in 2010.

In 2005–2010, Kunming's urban per capita green space showed an increasing trend. By the end of 2010 it reached 34 m², much higher than the national average of 10 m² per capita. Kunming was selected as one of the national Garden Cities. The forest coverage rate in Kunming reached 45.05%, 2.21 times as much as the national average.

5.4.5 *Low-Carbon Society*

Kunming's urban-rural income ratio during 2005–2010 was in the range of 2.95–3.25, while the national average was 3.22 over the same period. The income gap in Kunming is becoming larger.

Kunming's CO_2 emission per capita showed a notable increasing trend, having increased from 6.25 tCO_2/person in 2005 to 8.13 tCO_2/person in 2010. In the same period, the national average went from 4.34 to 5.67. Kunming's per capita CO_2 emission is 43.4% higher than the national average.

In terms of low-carbon management system, China has set explicit goals to achieve low-carbon transition. Pilot projects are conducted and the policy framework is being continuously improved. Kunming has established a Low-Carbon City Research Center, which is mainly responsible for studying and developing a low-carbon city development plan for Kunming and supervising the implementation of the plan.

5.5 Low-Carbon City Construction and Management

This section of the report gives an overview of Kunming's low-carbon city construction and management based on the LCCC supporting indicators and their action checklists. Assessment results show that Kunming has been actively promoting low-carbon city construction, and the work has been very fruitful.

5.5.1 *Urban Management*

Kunming has been vigorously promoting low-carbon city construction. On one hand, it established the Municipal Energy Conservation Leading Group with a clear management scope and responsibilities to guide low-carbon city management work; on the other hand, it developed and published a lot of policies and legal documents to regulate low-carbon city development. The policy framework is getting into good shape.

Kunming integrated the low-carbon concept into its planning. The government issued the Kunming Low-Carbon Economy Development Master Plan (2011–2020), setting the clear goal of reducing carbon dioxide emission per unit of GDP in 2015 by 35% relative to the 2005 level and 20% relative to the 2010 level; and by 2020, it should be reduced by more than 45% relative to the 2005 level. The plan also specified industrial policies, pricing policies and fiscal policies to support the achievement of this goal. The plan emphasized low-carbon development in key areas, and set goals to promote low-carbon transport and give priority to public transport development. It also aims to upgrade the resource recycling system and set low-carbon economy development on to a systematic, regulated and healthy track.

In terms of carbon management in urban areas, Kunming has not yet completed its carbon emission inventory or a good database. It is starting to compose a carbon emission inventory. The government has prepared a continuous work plan and assigned a dedicated team to create the inventory.

Kunming attaches great importance to resource recycling, establishing a recycling system, supporting recycling enterprises and taking the lead in creating a community-based trading-market-centered recycling network. Kunming organized various activities where residents can use their recyclables to exchange for flowers, daily necessities, tourism coupons or domestic services. This helped to increase the recycling rate of waste resources such as scrap metal, plastic, broken glass, paper, home appliances and other recyclables.

Kunming implemented a series of economic and fiscal policies for low-carbon development, including the Interim Management Guidelines on Special Funds for Energy Saving by the Finance Bureau and Industry and Information Commission, the Interim Management Guidelines on Sewage Charge Funds by the Environmental Protection Bureau, Finance Bureau and Audit Bureau, the Notice on the Adjustment of Energy-Saving Products in Government Procurement List by the Finance Bureau and DRC, and the Work Plan to Promote Energy Saving and New Energy Vehicles. These policies provide various tools to support low-carbon economic development including grants, subsidies, awards, subsidized loans, capital injection, government procurement of energy-saving products, etc. Earmarked support in Kunming for low-carbon development in 2011 increased by 16.34% from the 2010 level (Table 5.2).

Kunming issued the Agreement on Emission Reduction Targets with Public Institutions which explicitly requires municipal hospitals to have energy-saving lamps accounting for more than 85% of the total. Meanwhile, Kunming launched low-carbon publicity and education outreach activities, including preparation and dissemination of a low-carbon lifestyle guide that give tips on reducing one's carbon footprint in daily life (clothing, food, housing, transportation,

Table 5.2 Earmarked Support for Low-Carbon Development (2010).

Category	10,000 *yuan*	% in Government Budget
2010 Low-carbon expenditure		
Energy saving	11,444	0.33
Emission reduction	42,302	1.22
Environment protection	163,001	4.7
Low-carbon programs	7,493	0.22
Others	2,915	0.08
	3,462,884	—

Table 5.3   Number of Low-Carbon Demonstrative Projects.

	2001	2002	2003	2005	2006	2007	2008	2009	2010	2011	Total
Schools	1	26	14	17	54	52	52	45	46	54	361
Communities				10	26	33	16	15	26	22	148

Note: As there are no clear boundaries for the definition of low-carbon projects, all projects with low-carbon components were accounted for.

household appliances, etc.). Kunming implemented Energy Conservation in Families, Communities and Schools, a program that raises awareness and encourages public participation by initiating various campaigns and activities such as "Low-Carbon Families", "Low-Carbon Communities", "Low-Carbon Schools" and "Reducing Carbon Footprint", etc. At the end of 2011, Kunming had 148 low-carbon demonstration communities and 361 low-carbon demonstration schools (Table 5.3).

The government issued a few policies on green procurement, including the Notice on the Adjustment of Environmental Labeling Products in Government Procurement by the Finance Bureau and Environmental Protection Bureau, and the Notice on the Adjustment of Energy-Saving Products in Government Procurement List by the Finance Bureau and DRC. There is no information platform yet available on low-carbon planning and management documents or timely publicity of such documents.

Kunming has taken measures to strengthen its municipal infrastructure management. The Housing and Urban Rural Development Bureau created a gas supply management agency to regulate the source, quality, gas utilities and users. Meanwhile, Kunming Coal Gas Metering Station also monitors the coal gas produced and supplied by Kunming Coking Gas Co. Ltd. with a real-time on-line monitoring system. To meet the threshold for the China Habitat Environment Prize, Kunming plans to raise the gas penetration rate to 99.4%. In addition, the Kunming Gas Master Plan set specific policy measures to increase gas coverage. Kunming has also introduced the Implementation Plan for Accelerating Rainwater and Sewage Management and Urban Waste Resource Utilization (#KZF[2011] 53), and the Implementation Plan for Strengthening

Urban and Rural Municipal Garbage Collection, Transportation and Disposal (KZB[2012]14), Kunming Urban Water Conservation Regulations and other policies to improve water conservation, sewage treatment and infrastructure development. Specific targets are set that by the end of 2012, the eight districts in Kunming should have 800 regulated waste recycling outlets built, 1,000 new recycling trucks in operation and regulated waste recycling/sorting/collection teams established. At present, the waste disposal rate in Kunming has reached 100%, but there is no specific waste-associated energy management agencies or objectives established, or any specific institutions in charge of regulating waste separation and reduction.

5.5.2 *Urban Economy*

The urban economy refers to the regional clusters of industrial, commercial and other non-agricultural businesses. Urban economic development is an important material precondition for a city to function properly. A low-carbon economy features economic entities aiming to reduce greenhouse gas emissions with lower energy consumption and lower pollution. Low-carbon agriculture is an important part of a low-carbon economy. This section analyzes Kunming's low-carbon economy by looking into all three industries while focusing on the secondary industry and tertiary industry based on research and interviews.

Kunming attaches utmost importance to the low-carbon development of agriculture and rural areas. To develop and promote pollution-free green and organic agricultural products, Kunming has conducted soil testing for all arable land and provides technical guidance for proper fertilizer application. It has tried to reduce chemical fertilizer use by offering technical training, disseminating fertilization soil testing cards, introducing environmentally friendly fertilizers and advocating for comprehensive scientific farm management measures. Other agricultural energy conservation measures have been taken as well, such as introducing controlled/slow-release fertilizer, promoting new high-concentration fertilizer and new technologies for scientific fertilizer application, treating and utilizing agricultural waste, demonstrating IPM (Integrated Pest Management) pilot villages, introducing degradable plastic film, etc. Kunming also vigorously promotes the construction of biogas tanks in rural areas. It has developed technical specifications for biogas application. The target set in the master plan for biogas is that the penetration rate of biogas tank construction and upgrading in rural households exceed 19% by 2015 and 20% by 2020, which translates into installation of 30,000 new tanks by 2015 and 60,000 by 2020. Another target is to have more than 95% of biogas digesters become operational.

To support low-carbon transition of the industry, Kunming issued the Notice on Allocating 2012 Energy Saving Targets and Tasks, which set stricter market access policies for energy-intensive and polluting enterprises than the national policies. It is specified that the DRC and Industry and Information Committee are responsible for energy assessment, the Environmental Protection Agency is responsible for environmental impact assessment, while the DRC and Industry and Information Committee are responsible for project approval and documentation. Although there is no special fund for electricity demand side management, Kunming did develop a scheduled power consumption program which set priorities for power supply as "five assurances and four restrictions": ensuring supply for necessary domestic consumption, ensuring economic growth, ensuring energy reduction goals, ensuring profitability and ensuring priority projects, while restricting supply to businesses with high energy consumption, high pollution, low added value or failing to meet industrial policies and standards. Specific principles are to ensure power supply for residents, hospitals, schools, railway institutions, transportation hubs, water suppliers, gas suppliers, broadcasting agencies, financial institutions, telecommunications, agricultural production, oil production, transportation and other users related to public interest and national security.

Kunming also developed policy incentives to foster a number of large enterprises promising significant tax contribution, including incentives for investment, rewards for lower-level government that cultivated businesses, and earmarking fiscal support for phasing out obsolete production capacity (e.g. 650,000 *yuan* for 2010). These policies aim to foster and support large industrial enterprises with an annual turnover of over 1 billion *yuan*, 5 billion *yuan* and 10 billion *yuan*. Meanwhile, Kunming assigned the Industry and Information Committee to be responsible for the establishment and promotion of energy management systems among industrial enterprises. Annually four training and advising events are carried out on energy management systems; letters of responsibility for energy conservation targets are signed with key energy-consuming enterprises in Kunming; and the government will monitor and evaluate completion of energy-saving targets by interviewing and investigating key energy-consuming enterprises, industrial administrative departments and county governments. In 2010 the evaluation covered 30 businesses, 6 industrial authorities and 17 lower-level governments. Kunming's Twelfth Five-Year Plan of Industrial Development and Information Technology set the goal for the industry to achieve a total output of 15 billion *yuan* by 2015. The Kunming Energy-Saving Special Funds Management Guidelines stated that subsidized loans will be made available for renewable energy companies and energy-saving product manufacturers.

In addition, the industrial sector has specific renewable energy targets. Kunming has accumulated extensive practical experience and is in a leading position in solar thermal utilization. In 2011, Kunming also established a Hazardous Waste Treatment and Disposal Center, which is responsible for regulating industrial solid waste disposal and utilization and is required to write and publish work reports regularly. At present, Kunming's industrial solid waste comprehensive treatment rate has reached 41.33% (2010).

There are still weaknesses in Kunming's low-carbon economic development. There is the lack of regulation for non-CO_2 greenhouse gas emissions from industrial processes. Plans for phasing out obsolete production capacity are still to be developed for the Twelfth Five-Year Plan period. No energy monitoring system has been established in industrial enterprises. No energy management system has yet been constructed in any key energy-user companies.

Kunming's energy services industry development shows a good trend. Since 2008, Kunming has developed and implemented various incentives to encourage the registration and development of energy service companies. Financial incentives have been made available to encourage contracted energy management and CDM project services. As of 2010 two contracted energy management projects have received such financial support. Meanwhile, two workshops are organized annually to promote energy service among industrial enterprises. An active and regulated energy services market has been formed with both integrated energy service companies and specific technology-based energy companies providing energy audit and energy assessment services. Moreover, the Finance Bureau and Industry and Information Commission jointly issued the Interim Management Guidelines on Special Funds for Energy Conservation and Emission Reduction, providing subsidies to encourage the signing of voluntary emissions reduction agreements. Energy-saving and emission reduction tasks stated in the agreements have been carried out while the Municipal Energy Saving Business Leading Group is responsible for the promotion, publicity, evaluation and acceptance of energy-saving projects. Regular training courses on energy management systems are available for service institutions and businesses. Training courses include Introduction to the Energy Management System Standard (GB/T23331), Steps to Set up Energy Management Systems, Implementation of Energy Management Systems and Energy Management Systems Inspection and Improvement. The lecturers are from Kunming and Dezhou. Besides, regular experience-exchanging events are organized for public institutions regarding energy management. The Municipal Energy Saving Office is responsible for such training activities. Documentation on the trainings is available.

With the support of various policies, Kunming's strategically important low-carbon industries such as tourism, culture, bio-pharmaceuticals, new materials, etc., experienced rapid development during the Eleventh Five-Year Plan period, contributing to the remarkable achievements in industrial energy saving. However, the economic policies could not change the carbon-intensive nature of Kunming's industrial structure and energy-intensive industries still expanded rapidly. It is imperative for the parties concerned to develop a clear understanding of these issues and properly manage the dynamic interaction between rapid urbanization and the three industries as well as the associated energy consumption and carbon issues.

5.5.3 *Green Building*

Green buildings refer to buildings which can save resources (energy, land, water and materials) to the maximum extent during their whole building life cycle, and can help protect the environment and reduce pollution, provide people with healthy, convenient and efficient space, and can harmoniously coexist with nature. The development of green building is dependent on policy support.

Green building targets have already been incorporated into Kunming's master plan — the share of green buildings in all the newly constructed buildings should reach 10% in the downtown and 5% in the whole municipality by 2015, and further to 15% and 10% respectively by 2020. The government has actively introduced national policies including the MOHURD Twelfth Five-Year Plan of Green Buildings and Implementation Guides to Accelerate Green Building Development in China by the Ministry of Finance and MOHURD (MOF-MOHURD [167] to lower-level governments and developers in their daily work. Kunming HURD and Yunnan provincial HURD have organized seminars to raise awareness among decision-makers in the building sector every year. The upcoming Yunnan Green Building Evaluation Criteria takes into account energy, materials, land, water, and indoor and outdoor environments. By the end of September 2012, two projects in Kunming received the National Three-Star Green Building Design Award. Kunming pays close attention to building energy consumption statistics. It started to build a database in 2010 in accordance with the Comprehensive Energy Consumption Statistics Table for Urban Civil Buildings. It is planned that a data survey will be completed for newly constructed large public buildings by 2012 and an energy audit will be completed for all existing large public buildings by 2015. An energy audit and green building certification application will be carried out from 2013 onwards, and further become obligatory for all newly constructed large public buildings. Kunming also made its building energy consumption quota in accordance with national policies including Building Energy Efficiency Evaluation and Labeling Technical Guidelines, Building Energy Efficiency Evaluation and Labeling Management Guidelines, etc.

In terms of green building management, Kunming attaches great importance to renewable energy applications in buildings. In 2010, the MOHURD and Ministry of Finance approved Kunming as a Demonstration City for Renewable Energy Applications in Buildings, which required Kunming to make its short-term action plan. As a result, the Implementation Plan of Demonstration City for Renewable Energy Applications in Buildings (2010–2012) was developed, which required Kunming to complete a certain amount of building area with renewable energy demonstration and conduct capacity-building activities. Six million square meters were required to be completed by 2011. Since 2010, Kunming has issued the Decisions on Promoting Application and Management of Solar Water Heating Systems in Buildings (KunRenFa[2009]95), Solar Hot Water System Construction Drawings Design Depth, and Review Points on Solar Hot Water System Construction Drawings (KunJianTong [2010]396) to promote renewable energy applications through financial awards and subsidies. In 2010 Kunming HURD began to provide subsidies for demonstration projects in the city, mainly for solar energy application. Subsidy payments will be based on the merits of technical solutions. Evaluation is conducted by the HURD and Finance Bureau and results are publicized for public comments. Fifty percent of the subsidy will be paid if no serious objections are received during the public comment period. The Science and Technology Division of the HURD is responsible for building energy conservation work. Each year the HURD will organize lectures to raise awareness on green building. Green building professionals also participate in Green Building Rating Standards training seminars organized by the Yunnan provincial HURD.

Green building work in Kunming has shown good effects. However, there are also challenges. The majority of residents, though they would be the major beneficiary of green building development, are not very aware of the issue. Moreover, although green buildings are economically very worthwhile from the life-cycle perspective, their payback time is rather long. Building management has a large impact on energy saving, yet property management companies generally have very low incentives or capacity to be committed to good energy management. Therefore, policies should be developed to involve and motivate all stakeholders including designers, contractors, property management companies, real estate agents, material suppliers and house owners. It will be easier to promote green building if all parties reach a consensus at the beginning.

5.5.4 *Low-Carbon Transport*

The transportation system is an integral part of urban construction and urban modernization. Low-carbon transport refers to a transport system featuring high energy efficiency, low energy consumption, low pollution and low emissions. It can be achieved through increasing energy efficiency, upgrading the energy mix and improving the development model. Kunming explored options for building a low-carbon transport system according to local conditions.

Kunming prepared the Low-Carbon Transport Development Plan (2011–2020), which contains clear objectives: taking low-carbon development as the guiding principle, relying on the transformation of the modes of transportation and development, and reducing carbon emission in each process through technological innovation, energy mix upgrading, and managerial improvement and change of consumption patterns. Comprehensive policies and regulations will be developed to support low-carbon transport development; a carbon emissions database and a monitoring and evaluation system will be basically established; and the overall ability of the transport sector to address climate change will be significantly enhanced. By 2020, a low-carbon transport system will be developed in line with national requirements, with Kunming characteristics. With this plan developed, however, there is no transport energy and emission inventory established and the integrated traffic management measures are missing to date.

Kunming vigorously promotes the development of its public transportation system and was the first city to build BRT in the country. Kunming released several policies including the Kunming Public Transport Development Plan, Kunming Urban Public Transportation Network Layout Plan (awaiting approval) and Decision of Kunming CPC Commission and Kunming People's Government on Accelerating the Development of Urban and Rural Public Transportation. Kunming Urban Transport Research Institute and Kunming Urban Social and Economic Investigation Team jointly completed the *Kunming Urban Transportation Development Annual Report* (2010) based on a survey that investigated the average commuting time of the residents. The municipality made its budget available to subsidize bus fares, fuel, vehicle insurance and car purchase loans. Taking the opportunity that the Ministry of Transport had selected it as a pilot city to apply the taxi intelligent service system, Kunming promoted the construction of intelligent transportation, focusing on developing a modern passenger information system, constructing a passenger public information service platform, establishing an intelligent taxi scheduling management information system, building a bus travel information service platform, applying a dynamic bus monitoring system and exploring an integrated transport intelligent platform. Kunming attaches great importance to the use of new energy vehicles. It has issued the Notice on Implementing Energy Efficient and New Energy Vehicles Demonstration Pilot Project (FB-HURD[2009]6), Implementation Plan for Energy Efficient and New Energy Vehicles Demonstration Pilot Project, and Financial Subsidies Management Guidelines for Energy Efficient and New Energy Vehicles Demonstration Pilot Project. This project set the goal of having 1,000 new energy vehicles put into operation during 2009–2012. Financial incentives are provided using both the central government budget and the municipal budget. Beginning in 2010, new energy infrastructures such as gas stations and car recharging stations are being constructed along the main roads. By the first half of 2012, 750 new energy vehicles were put to use, accounting for 60% of all incremental vehicles. In addition, Kunming held regularly awareness-raising campaigns adding to the theme of the year (Harmonious

Kunming, Green Homes, Everyone Participating in 2006; Protecting Colorful Yunnan, Constructing Ecological Kunming in 2007; Green Olympics, Colorful Yunnan, Ecological Kunming in 2008; Protecting Dianchi Lake, Constructing Ecological Kunming in 2009; and Low-Carbon Emission, Ecological Kunming in 2010). The number of participants increased from 500 in 2005 to 600 in 2010.

5.6 Challenges

From the LCCC Primary Indicator System evaluation results, it can be concluded that Kunming has made some achievements in its low-carbon city construction. Its carbon productivity, carbon intensity of energy mix and share of non-commercial renewable energy use all exceeded the national average. Its economic growth rate is higher than its carbon emission growth rate, indicating that carbon emission and economic growth are getting relatively decoupled. Kunming's environment indicators are significantly better than average. However, there are still difficulties and challenges.

Judging from the assessment, Kunming has very good results in many indicators in the economy, infrastructure and environment. However, its energy intensity and carbon emissions per capita are slightly higher than the national average. There is much room for improvement in terms of energy mix and energy efficiency.

Based on the assessment of its efforts, its performance in indicators in energy, society and facilities is not ideal and faces enormous challenges for improvement. Evaluation of its efforts is mainly based on the degree of improvement. As Kunming is currently doing much better than the nation as a whole, it will be more difficult to improve further.

(1) Low Energy Efficiency and Management Level

The key reason for this is that Kunming's industrial development is relatively backward. Kunming has more than two-thirds of the province's key industries and key enterprises, but they are mostly resource-dependent and energy-intensive ones with low energy efficiency and extensive growth patterns. Hence it is contributing to a large proportion of the province's total economic output as well as total energy consumption. Kunming is located in a relatively underdeveloped region, whose overall economic strength is not high and technology development is not very advanced. Kunming's six pillar industries — tobacco, chemical, pharmaceutical, metallurgy, machinery and power generation — are all energy-intensive industries with limited willingness to sign voluntary energy conservation agreements. There is no complete energy management system or energy inspection system.

(2) Carbon-Intensive Industrial Structure

The nine energy-intensive industries — electricity, heat production and supply; ferrous metal smelting and rolling, chemical materials and chemical products manufacturing, petroleum processing and coking and nuclear fuel processing, gas production and supply, non-metallic mineral products, non-ferrous metal smelting and rolling processing industry, non-metallic mining industry, ferrous metal mining — together account for more than 80% of the total energy-consumption-related carbon emissions by all industries. This is a tremendous challenge for Kunming to reduce its carbon intensity.

(3) Difficulties Accelerating Renewable Energy Development

Currently Kunming is still in the phase of industrialization and rapid urbanization, when energy demand will increase with the rapid economic development. As a result, energy-related carbon emissions will inevitably continue to grow.

Developing solar, biomass and other renewable energy is required for the construction of an ecological and low-carbon economy. Currently renewable energy development in Kunming has several problems. First, the lack of integrated and target-binding planning and the incomplete local regulation safeguards make high penetration difficult to achieve. Second, policy support is still insufficient especially for the initial stage of renewable energy development and utilization. Government support is urgently needed for the formation of a local brand and market. Third, the capability for independent innovation and technology development needs to be strengthened.

5.7 Conclusion and Recommendations

Kunming's low-carbon city construction has several highlights. It has a mild climate which results in low heating and cooling energy demand. It is rich in vegetation and forest, hydro, solar and other renewable energy. One of the highlights is the broad renewable energy development and utilization, and another is the development of low-carbon industries such as tourism and its flower industry based on regional advantages. Moreover, its environmental remediation and protection work are also remarkable.

The Twelfth Five-Year Plan period is critical for the city's transition to a low-carbon economy. First, Kunming has significant motivation for economic development; second, the municipality proposed the development of a low-carbon eco-city, which is an inherent demand and is in line with the fundamental interests of the people of Kunming; third, in recent years its development in the renewable energy industry has provided Kunming with a new growth point for the transition from the traditional high energy consumption city to a sustainable low-carbon city; fourth, such low-carbon efforts also echo China's strategy of scientific development and building a harmonious society.

(1) Improving the Low-Carbon Development Management System

Recently, Kunming formulated a series of policies to implement its low-carbon development policy, developed the Low-Carbon Development Plan, and carried out research on current carbon emissions and future emission reduction potentials. Necessary administrative and research institutions were established, which has laid a good basis for carrying out research and policy-making. In addition, Kunming endeavors to develop low-carbon industries including its two pillars of tobacco and tourism as well as those industries in Kunming's Top Ten Industrial Revitalization Program: tourism, culture, trade and logistics, headquartered economy, new energy, optoelectronics, biotechnology industry, etc. Kunming has accumulated certain research and development capacity and a certain scale of production for several industries, including solar photovoltaics, solar thermal, biomass energy development and utilization. This also helps provide a good industrial base for its low-carbon development.

Kunming further strengthened its government management mechanism and established management institutions with clear responsibilities and tasks for low-carbon city development. The main objective of such an institution is to coordinate the two important goals of economic development and energy saving. Take planning as the starting point — integrate the low-carbon concept and measures into the planning of all sectors, and provide effective tools to monitor energy consumption and carbon emissions for the implementation of the plans. Conduct annual reviews in order to avoid and to correct mistakes. To reduce the pressure of reaching energy conservation targets, efforts need to be made for each phase of the cycle.

(2) Compiling a Low-Carbon Inventory and Developing Low-Carbon City Management Tools as the Basis for Policy-Making

According to the LCCC Indicator System recommendations, the tool should include three databases or software: first, an urban carbon emission inventory which is a top-down model established with statistics collected and used for monitoring and forecasting the trend of different sectors; second, energy consumption data on different types of buildings; third, an emission inventory of different modes of transportation, including buses, private cars, taxis and so on. In compiling the emission inventories, Kunming should closely coordinate with Yunnan provincial government and the central government so its inventories will be consistent with national and provincial inventories. A database development and updating mechanism should be established as well to provide the basis for Kunming's climate policy development.

(3) Upgrading the Industrial Structure and Promoting Low-Carbon Industries

Vigorously developing industries and low-carbon development are not necessarily incompatible with each other. An excellent environment will make the city more attractive to capital and industries, while economic development will also accumulate capital for the investment of low-carbon development projects. In the process of industrial development Yunnan should accelerate the elimination of backward production capacity, and strictly control market access criteria for energy-intensive and highly polluting industries. It is recommended to promote energy management systems in the industrial sector and further develop and disseminate energy management systems among service companies, public institutions as well as in the construction and transportation sectors. More support should be given to promote the development of carbon-neutral tourism, modern service industry, urban low-carbon agriculture and low emission industries.

(4) Further Developing Non-Commercial Renewable Energy

Enlisted as one of the National Low-Carbon Pilot Cities, Kunming has a greater chance of getting major national energy projects invested in Kunming, which will help promote local non-commercial renewable energy development and utilization, and help increase the proportion of renewable energy consumption and upgrade its energy mix and energy consumption patterns. Nationwide, the share of non-commercial renewable energy (mainly biogas and solar thermal) is approximately 1.1% in the total energy consumption (2010). Kunming has very favorable conditions for developing non-commercial renewable energy utilization. It should actively pursue non-commercial renewable energy utilization through supporting the development of renewable energy industries and fostering the market.

At the same time, synergetic effects can be achieved during the low-carbon city construction transformation. For example, the investment on low-carbon projects will also create new employment opportunities. Human resource development is also important in non-commercial renewable energy research and development and the renewable energy base construction and should get sufficient attention.

(5) Strengthening International Cooperation to Get External Funding and Technical Guidance and Support

Kunming has a unique geographical advantage, a huge potential to attract foreign investment. More investment can also boost the logistics industry and create other new economic growth points. In addition, cooperation projects usually bring in advanced knowledge and successful experience of other regions. Developed European countries have made remarkable achievements in their low-carbon city construction and have accumulated a wealth of low-carbon city construction technologies, methodologies and experience. Kunming can pursue further international cooperation based on the LCCC project.

References

1. Institute for Urban and Environmental Studies in CASS, LCCC Project Management Office. 2012. *China Low-Carbon City Evaluation System Methodology Report.*
2. Institute for Urban and Environmental Studies in CASS, LCCC Project Management Office. 2012. *China Low-Carbon City Evaluation System — Primary Indicators and Supporting Indicators.*
3. Kunming Development and Reform Commission. 2010. *Kunming Low-Carbon Economy Development Master Plan* (2011–2020).
4. Kunming Transportation Bureau. 2010. *Kunming Transportation Low-Carbon Development Plan* (2011–2020).
5. Kunming Development and Reform Commission. 2012. *Implementation Scheme of Kunming Low-Carbon Pilot City Program.*
6. Kunming Institute of Environmental Science, Kunming Low-Carbon City Development Research Center. 2012. *Kunming Low-Carbon City Evaluation Index System Investigative Report.*
7. Beijing Energy and Environment Company. 2012. *Energy Conservation Strategies and Action Plans for Kunming's Five Major Energy Consuming Industries.*
8. Beijing Energy and Environment Company. 2011. *Energy Consumption and Energy Efficiency of Kunming's Five Major Energy Consuming Industries.*

Appendix: Recommended Low-Carbon Action Plan for Kunming

#	Recommended Action	Details	Responsible Dept.	Priority	Deadline (years)
I.	**Urban management**				
1.	Preparation of comprehensive carbon emissions inventory of the city	Establish emission monitoring, statistical, feedback and long-term regulatory mechanisms. Compile emissions inventory based on statistics system	DRC	High	2
2.	Develop emissions inventory preparation programs and capacity-building programs	Responsible departments issue specific program, and invite experts to comment	DRC	High	0
3.	Establish evaluation mechanisms for the effect of funds allocated	Establish feedback channels and periodic verification of funds, compare with expected results	FB leading, other institutions supporting	Low	0.5
4.	Integrate green procurement into government planning	Develop Green Product Directory based on green procurement list to regulate procurement	Government Purchase Office	High	1
5.	Prepare green products directory and supporting files	Prepare the directory based on principles such as energy/water saving, low pollution, low toxicity, renewable and recyclable with reference to the country's list	Government Purchase Office	High	1

(*Continued*)

(*Continued*)

#	Recommended Action	Details	Responsible Dept.	Priority	Deadline (years)
6.	Establish information platform for transparency of low-carbon planning and management information	Establish information platform via the Internet, radio, newspapers, etc., to publicize low-carbon planning and management information	DRC	High	0.5
7.	Collect and publicize low-carbon-related information including policies, initiatives, projects, activities, etc.	Collect and publicize low-carbon-related information including policies, initiatives, projects, activities, etc.	DRC	High	1
8.	Build channels for feedback on low-carbon planning and management	Through newspapers, administrative agencies and other communication channels	DRC	Medium	0.5
9.	Collect comprehensive data on renewable energy production	Collect comprehensive data on renewable energy production	HURD, DRC, SB	High	1
10.	Set clear water-related energy management targets	Set feasible targets through surveys and discussions	Water Bureau	Medium	0.5
11.	Set clear regulations for sewage treatment plant energy management	Responsible department issues explicit regulation guidelines in accordance with local actualities	Dianchi Management Bureau	High	1
12.	Set clear waste-water-associated energy management targets	Set feasible targets and integrate them into urban planning	Dianchi Management Bureau	High	1

(*Continued*)

(Continued)

#	Recommended Action	Details	Responsible Dept.	Priority	Deadline (years)
13.	Develop water conservation action plan	Responsible departments set targets and action plans respectively for residential and industrial water use reduction	Residential: Public Utilities Bureau; industrial: IIC	Medium	1
II.	**Green economy**				
14.	Get GHG emission statistics for industrial processes	Establish mechanisms for emission monitoring, statistics collection, feedback and long-term regulation of industrial enterprises. Build emissions inventory accordingly	IIC, SB, DRC	High	2
15.	Develop systematic DSM planning	Develop systematic DSM planning from consumption perspective	IIC — Energy Division	High	1
16.	Promptly publicize energy monitoring results	Promptly publicize energy monitoring results	IIC — Energy saving Management Division	Medium	0.5
17.	Set specific targets for renewable energy use among industries	Set specific targets for renewable energy use among industries based on urban master plan	DRC, IIC — Energy saving Management Division	High	1

(Continued)

(Continued)

#	Recommended Action	Details	Responsible Dept.	Priority	Deadline (years)
18.	Regularly write and publish reports on industrial solid waste utilization	Set special projects and funds to write and publish reports on industrial solid waste utilization	EPB, IIC — Resource Utilization Division	Medium	1
19.	Appoint full-time energy manager with clear responsibilities in public institutions	Leading institution should supervise that at each responsible departments, a full-time energy manager with clear responsibilities is appointed	Government Service Bureau, Business Bureau, HURD	High	1
20.	Sign voluntary emissions reduction agreements with service companies to promote low-carbon consumption	Develop template, conduct publicity campaign and encourage service companies to sign voluntary emissions reduction agreements	Business Bureau, Tourism Bureau	High	2
21.	Provide policy incentives to encourage energy service companies to register and operate	Provide (both material and non-material) incentives to encourage energy service companies to register and operate	IIC	Medium	1
22.	Provide relevant information and technical support for producers regarding agriculture and forestry residue treatment and reuse	Provide relevant information (including manuals, notices and documents) and technical support for producers regarding agriculture and forestry residue treatment and reuse	AB — Eco & Environment Protection Division, Forestry Bureau	High	2

(Continued)

(Continued)

#	Recommended Action	Details	Responsible Dept.	Priority	Deadline (years)
23.	Issue policies that encourage farmers to get certified for green/organic/pollution-free products	Issue policies that encourage farmers to get certified for green/organic/pollution-free products	AB	High	1
24.	Develop a plan to achieve agricultural energy savings	Responsible department develops a plan with specific measures	AB — Rural Energy Office	High	2
III.	***Green building***				
25.	Develop a sound work plan to collect building energy statistics (including the statistical classification for all building types)	Establish mechanisms for emission monitoring, statistical, feedback and long-term regulation. Compile emissions inventory based on statistics	HURD — Policy and Regulation Division	High	3
26.	Prepare energy use plan and action guidelines for new buildings in new urban area	Prepare energy use plan and action guidelines for new buildings in new urban area	HURD — Urban Construction Division	High	1
27.	Preparation of new urban planning, new buildings, energy use, planning, action guidelines	Further promote building integration retrofit for solar and other renewable energy. Organize campaigns and provide policy and financial incentives	HURD — Tech & Info Division	High	3

(Continued)

(Continued)

#	Recommended Action	Details	Responsible Dept.	Priority	Deadline (years)
28.	Integrate renewable energy applications into administrative regulatory review process (except drawing inspection phase)	Integrate renewable energy applications into administrative regulatory review process	HURD — Tech & Info Division	High	2
IV.	**Low-carbon transport**				
29.	Start preparing carbon emissions inventory for the transport sector	Establish emission monitoring, statistical, feedback and long-term regulatory mechanisms. Compile emissions inventory based on statistics system	TB — Integrated Plan Division	High	2
30.	Clarify funding sources and the responsible departments to implement integrated transport	Clarify funding sources and the responsible departments to implement integrated transport	TB — Integrated Plan Division	High	0.5
31.	Train procurement staff on new energy vehicles	Conduct regular training activities	TB — Policy & Regulation Division	Medium	2
32.	Collect data on the number of existing road lamps and their energy consumption	Collect data on the number of existing road lamps and their energy consumption	Law Enforcement Bureau — Urban Lighting Division	High	1

(Continued)

(Continued)

#	Recommended Action	Details	Responsible Dept.	Priority	Deadline (years)
33.	Formulate guidelines and action plan that gives priority to public transport	Responsible departments take the lead in formulating guidelines and action plan that gives priority to public transport	TB and bus companies	Medium	1
34.	Set specific measures to increase public transport service, including punctuality, convenience, comfort level and affordability	Conduct surveys and expert seminars to develop such specific measures	TB and bus companies	High	1
35.	Develop integrated plan for slow traffic key infrastructures	Develop integrated plan for slow traffic key infrastructures	PB	High	1

Abbreviations

DRC: Development and Reform Commission
HURD: Bureau of Housing, Urban and Rural Development
IIC: Industrial and Information Commission
FB: Financial Bureau
PB: Planning Bureau
AB: Agriculture Bureau
TB: Transport Bureau
BERO: Building Envelop Retrofit Office
BEEST: Building Energy Efficiency Science and Technology Department Energy Monitoring Committee, Public Building Energy Management Agency
TMDSTD: Traffic Management Division Safety and Technology Department

Chapter Six

Assessment of Baoding's
Low-Carbon Development

6.1 Facts of Baoding City

Baoding is located east of the Taihang Mountains, in the western Hebei plain, and in the middle-west of Hebei Province. Baoding's jurisdiction consists of three districts (South District, North District and New District), four cities (Dingzhou City, Zhuozhou City, Anguo City, Gaobeidian City), eighteen counties (Yixian, Xushui, Laiyuan, Dingxing, Shunping, Tang, Wangdu, Laishui, Qingyuan, Mancheng, Gaoyang, Anxin, Xiong, Rongcheng, Quyang, Fuping, Boye and Li County), one High-Tech Industrial Development Zone and the Baigou New City. Baoding covers a total area of 22,190 km^2 and has a population of 11.20 million (2010).

Baoding is the gateway to Beijing. Geographically it is the connecting region of many cities and provinces. It was the capital city in several dynasties in China's history. The city has a rich cultural heritage, including 47 national key cultural relics and 17 national intangible cultural heritages. Currently it is one of the two wings of the Large Beijing Economic Circle and one of Beijing's major satellite cities.

From the geographical point of view, Baoding is located in the heart of the Beijing-Tianjin-Shijiazhuang triangle (Figure 6.1). The urban center of Baoding is 140 km south of Beijing, 145 km west of Tianjin and 125 km northeast of Shijiazhuang. It has direct access to Beijing Capital airport, Zhengding Airport and the ports in Tianjin, Qinhuangdao and Huanghua. It has an extensive transport network with many major railways and roads passing its territory, including Beijing-Guangzhou railway, Beijing-Kunming highway, 107 State

Figure 6.1   Baoding Jurisdiction.

Road, Beijing-Hong Kong-Macao highway, Baoding-Tianjin high-
way, Binhai-Baoding highway, Baoding-Cangzhou highway, Zhang
Jiakou-Shijiazhuang highway. Four highways — Beijing-Hong Kong-
Macao highway, Rongcheng-Wuhai highway, Beijing-Kunming high-
way and Baocang-Baofu highway — connect at peripheral areas of
Baoding, forming a network of highways around the city.

Baoding covers an area that increases northwestward in elevation.
It features a mountainous region and a plain, both of which are rich
in natural resources. The mountainous region has over 50 kinds of
mineral resources including copper, iron, coal, zinc, aluminum, gold,
silver, molybdenum, asbestos, mica, limestone, bauxite, marble, etc.
Laiyuan County has the third largest molybdenum reserves in the
country. The variety and quantity of mines in Laiyuan is the largest in
Baoding. Quyang has large and high-grade marble and bauxite
reserves. Baiyangdian plain has abundant oil, natural gas and geothermal resources with great use potential. This area is also home to seven
national grain production bases, eight fruit-growing bases and eight
animal husbandry and fishery bases.

6.2 Practice and Highlights

As one of the first batch of low-carbon pilot cities selected by the World Wildlife Fund (WWF) and NDRC, Baoding has launched a variety of initiatives to explore the low-carbon development transformation on the basis of its own advantages. In 2006, Baoding proposed its strategic vision of Baoding as China's Electricity Valley, focusing on the development of wind power and solar photovoltaic industries. It has attracted and fostered more than 170 new energy companies in the region. In 2007, Baoding raised the standard of becoming a Solar Energy City. Solar PVs are widely installed in public places, residential blocks and tourist attractions, making substantial contribution to its transition to a low-carbon city. Because of its efforts and achievements, Baoding was selected by the WWF as a pilot city for the China Low-Carbon City Development Project in January 2008. Taking these opportunities, Baoding gradually figured out its roadmap for transition to a low-carbon economy — led by low-carbon industries and featuring a low-carbon lifestyle.

For the construction of a low-carbon city, Baoding takes changing industrial growth patterns and decarbonization of industries as the first priority. On the one hand, it endeavors to guide new energy and energy equipment industries to form their own technological and industrial advantages. From the late 1990s, Baoding began to explore the development of new energy industries. Large programs including China Electricity Valley and Solar Energy City were started. As the brand of China Electricity Valley expands and the industrial cluster rapidly grows, some of the world's top 500 enterprises have came to Baoding to invest in new energy industries. Hyosung Corporation from South Korea established a joint venture with Baoding Tianwei — Baoding Hyosung Tianwei Transformer Co., Ltd. Mitsubishi Electric set up Mitsubishi Tianwei Transmission Equipment and Transformer Co., Ltd. with Tianwei Group. France Air Liquid Company also invested in new energy industries in Baoding. In addition, large domestic enterprises committed to new energy industries such as China South Industries Group and China Guodian Corporation have also started a series of large projects. Currently, Baoding has a number of large new energy companies, such as Tianwei Yingli — the only company with a full production industrial chain for polycrystalline silicon solar cell components and systems, and AVIC Huiteng Wind Power Equipment Co., the nation's largest manufacturer of wind turbine blades, the only one with self-owned intellectual property rights and R&D capabilities for wind turbine blades. According to Baoding Development and Reform Commission, the total turnover of the new energy industry was only several billions of dollars in 2006; however, it increased to more than 10 billion *yuan* by the end of 2007, 24.45 billion *yuan* in 2008, 31.8 billion *yuan* by the end of 2009 and 44 billion *yuan* in 2010. The advantages of the new energy industry cluster have emerged. Baoding has cultivated an industrial cluster with its core industries of solar photovoltaics and wind power and supporting industries of new energy storage materials, power transmission equipment, power electronics and automation, high efficiency energy-saving equipment and biomass business.

On the other hand, Baoding aims to decarbonize its local traditional agriculture. Agriculture is both foundational to the national economy and a major source of greenhouse gas emissions. Agricultural greenhouse gas emissions are mainly from farming and aquaculture production process and waste — especially the residue and the excessive application of chemical fertilizers and pesticides. Low energy consumption and low-carbon agriculture refer to agricultural production process featuring low pollution, low emissions and large carbon sinks, realized through minimizing the use of chemical fertilizers, pesticides, machinery, increasing the area of forest and vegetation, changing irrational production patterns and lifestyles in order to reduce greenhouse gas emissions, and aiming to achieve "high efficiency, high quality, high yield". Agriculture is the mainstay industry of Baoding, which comprises of 22 counties and 5 districts. The rural area is much larger than the urban area in Baoding. Baoding government takes water, fertilizer, pesticide, seeds, land and energy as its six focus sectors, trying to promote the transition to low-carbon agriculture through the implementation of the Ecological Homes project, straw utilization projects and agricultural water-saving projects. Ecological Homes is a program initiated by the Ministry of Agriculture with the goal of changing the traditional family-based production and lifestyle and integrating the previously family-based technologies and practices. Rice straw utilization is one of the major components in circular agricultural economy. During the past few years, Baoding has vigorously promoted comprehensive utilization of straw and straw-based technologies, including straw compression as feeds, whole corn stalk preservation technology and stalk composting. Meanwhile, some crop-based agriculture byproducts are used for mushroom production as well, turning waste into valuable resources and increasing farmers' income. In 2011, the total biogas produced was 13,400 m^3 in Baoding, representing every saving of 95,600 tons of coal equivalent. The rate of comprehensive utilization of straw has increased to 86%. Irrigation efficiency of wheat and vegetables improved too, resulting in water saving of 280 million m^3. In summary, Baoding has made notable achievement in low-carbon agriculture.

Baoding seeks to transition to low-carbon economy through projects that upgrade municipal infrastructure and transportation network, applying renewable energy at large scale and guiding people to adopt low-carbon lifestyle. Take the transport sector for example. As Baoding is identified as one of the first ten low-carbon transport pilot cities by the Ministry of Transport, it initiated the gasification program and achieved remarkable results. In 2012, Baoding purchased 689 units of natural gas buses to replace all non-gas buses, put 260 LNG vehicles in operation and had 82.3% of the city's taxis shifted from fuel to gas. In the urban area six gas recharge stations were built and another four are under construction. The awareness of low-carbon transport of Baoding residents increases as they travel in such vehicles. At the same time, demonstrative projects at businesses, parks, communities and villages also play a role in increasing people's awareness. Power Valley International Hotel is another example of increasing awareness through projects. It is seen as a lighthouse project among Baoding's low-carbon buildings. The hotel is the first eco-friendly hotel with solar power generation connected to the grid in China, a showcase of integrating photovoltaics with building construction. It has become the landmark building of Baoding's transition to low-carbon development. Residents who live around or pass by the building will get a chance to learn about how renewable energy brings convenience to life and be inspired.

Baoding's success in low-carbon city development comes from several drivers. First, Baoding has a strong urge for economic expansion; second, the Baoding government proposed the development of a low-carbon eco-city based on its inherent demand and the fundamental interests of the Baoding people; third, its recent strategy of developing the renewable energy industry and the Sun City goal provided Baoding a new growth point in its transition from a traditional energy-intensive city to a sustainable low-carbon city; fourth, such work fit with China's national strategy of pursuing scientific development and building a harmonious society.

6.3 Status Quo and Efforts

Baoding is one of the pilot cities in the Sino-Swiss partnership project, the Low-Carbon City in China project (LCCC). Based on the LCCC Primary Indicator System and related logical framework, the report evaluates Dezhou's low-carbon development from two perspectives: status and level of effort. Major data sources include: (1) a variety of publicly available statistic data, government documents and statistical bulletins; (2) field research and interviews with relevant departments and stakeholders; (3) domestic and international relevant low-carbon city index system research results.

The LCCC Primary Indicator System includes 15 indicators in 5 categories: economy, energy, infrastructure, environment and society. Among them, the economic indicators reflect a city's low-carbon economic development stage and implementation of the country's overall energy reduction targets; low-carbon energy indicators reflect regional low-carbon energy resources as well as the city's effort in upgrading its energy mix; infrastructure indicators show the urban infrastructure level and low-carbon consumption levels. Environmental indicators assess a city's green development. Indicators on society evaluate consumption patterns and social equity. Specific evaluation results are shown in Table 6.1.

Table 6.1 Low-Carbon Development Assessment of Baoding.

Nos.	Indicators	Unit	Baoding 2005	Baoding 2010	China 2005	China 2010	Effort
	Economy						
(1)	Carbon productivity	10,000 *yuan*/ton CO_2	0.325	0.413	0.305	0.424	100
(2)	Energy intensity	tce/10,000 *yuan*	1.276	1.033	1.221	0.974	80
(3)	Decoupling index	—	0.92		0.88		80
	Energy						
(4)	Non-fossil energy in primary energy consumption	%	6.8	8.6	0.2	0.8	100
(5)	Per capita non-commercial renewable energy use	kgce/per capita	16.5	26.7	33.9	53.4	100
(6)	Carbon intensity of energy	ton CO_2/tce	2.405	2.341	2.69	2.66	15
	Infrastructure						
(7)	Energy consumption per unit building area for public buildings	kgce/m^2	27.3	23.86	26.83	13.42	100
(8)	Energy consumption per building area for residential buildings	kgce/m^2	13.3	11.9	12.05	6.68	100
(9)	Ratio of green transport	Standard units	7.57	9.2	5.99	12.3	100
	Environment						
(10)	Percentage of days with API less than 100	%	51.9	81.7	84.38	90.68	80
(11)	Forest coverage rate	%	149.80	105.04	71.05	96.63	88
(12)	Domestic water consumption per capita per day	kg	18.21	20.36	17.54	18.86	75
	Society						
(13)	Urban-rural income ratio	%	3.22	3.22	2.500	2.763	80
(14)	Per capita CO_2 emission	ton CO_2/per capita	4.34	5.67	3.219	4.312	100
(15)	Low-carbon management institution	—	No	Rather good	No	Good	100

Note: Data on public building and residential building energy consumption is from sample survey results of the HURD.

6.3.1   *Low-Carbon Economy*

In accordance with the evaluation by the LCCC Primary Indicator System, Baoding's carbon productivity in 2010 was 0.424 per 10,000 *yuan* GDP/tCO_2, an increase of 39.01% compared with the 2005 level. Compared to the national average of 0.413, it is 3.41% higher. This shows that carbon productivity in Baoding is rather high, and forms a good basis for low-carbon development.

However, Baoding's economic structure is industry-dominated (industries accounting for 54.9% in the economy in 2011), while energy-intensive heavy industries accounted for a very large proportion of all industries (industries contributed to 72.5% of total energy consumption and "industries above a certain scale" contributed 56.8% of total energy consumption). The eight major energy-intensive industries (textiles, paper and paper products, petroleum processing, coking and nuclear fuel processing, chemical materials and chemical products manufacturing, non-metallic mineral products, ferrous metal smelting and rolling processing, transportation equipment manufacturing, electrical machinery and equipment manufacturing) contributed to 77.8% of total industrial energy consumption. Continuous development of energy-intensive industries will require a large amount of fossil energy supply and increase the pressure on carbon emission reduction. Baoding's industrial structure has long been investment-driven, energy-intensive and pollution-intensive, which poses great challenges to sustainable development.

During 2005–2010, energy consumption in Baoding increased year on year, with energy consumption per unit of GDP showing a gradually declining trend — a decrease rate of 4.05% per year and a total decline of 20.23%, a higher rate than the national average of 19.1%. This shows its energy consumption was declining relative to its economic development, indicating an improvement. However, in terms of absolute energy consumption per unit of GDP, Baoding's data was 0.974 tce/10,000 *yuan* GDP in 2010, 5.69% lower than the national average. Baoding made certain achievements in reducing its unit GDP energy consumption and upgrading its energy consumption structure and industrial structure. Baoding's unit GDP energy consumption is lower than the average level of Hebei Province and of the country too. However, based on the assessment, the level of its effort was only 80%.

During 2005–2010, the country's carbon decoupling index was 0.92, while it was 0.88 in Baoding, both showing a relative decoupling trend.

6.3.2 *Low-Carbon Energy*

Baoding has a coal-dominated energy mix. In 2005–2010, Baoding's share of non-fossil fuels in primary energy consumption rose from 0.2% to 0.8%, but still lags far behind the 2010 national average of 8.6%. In terms of non-commercial renewable energy consumption per capita, during 2005–2010, Baoding went from 33.9 kgce to 53.4 kgce, an increase of 56.3%. In the same period the national average went from 16.5 kgce to 26.7 kgce, an increase of 61.82%. Baoding is apparently doing better than the average level of China in this respect.

Taking advantage of its renewable energy industries, Baoding started to promote large-scale application of solar facilities since 2007, and became a city with high penetration of solar facilities in China and in the world. By September 2009, Baoding completed its project of replacing street light lamps with solar energy lamps along the main roads, which results in electricity saving of at least 19 million kWh every year. Meanwhile, Baoding constructed solar power plants in its jurisdiction, which demonstrates the application of photovoltaic products.

During 2005–2010, Baoding's carbon intensity of energy dropped from 2.69 tCO_2/tce down to 2.66 tCO_2/tce. As Baoding's proportion of coal consumption was significantly higher than the national average, its carbon intensity of energy consumption was 13.68% higher than the national average of 2.341 tCO_2/tce. This has to do with its fossil-fuel-based energy mix.

6.3.3 *Low-Carbon Infrastructure*

During 2005–2010, Baoding's public building energy consumption per unit area reduced from 26.83 kgce to 13.42 kgce, a decline of 50%. During the same period the national average went from 27.3 kgce to 23.86 kgce, a 12.6% decline. Absolute energy consumption is lower in Baoding than the national average, while its reduction rate is higher than the national average.

For residential buildings at the same time, Baoding's energy consumption per unit area reduced from 12.05 kgce to 6.68 kgce — a decline of 44.6%, while the national average went from 13.34 kgce to 11.9 kgce, a decline of 10.79%. Judging from both absolute consumption and the reduction rate, Baoding has made notable progress.

In 2010, the green travel rate has reached 70% in Baoding. The number of buses per 10,000 people increased to 12.3 standard units, far higher than the national average of 9.2 standard units. Such achievements were a result of the integrated low-carbon transportation program, which aimed at promoting green modes of travel, including walking, cycling and high efficiency low-carbon public transport, reducing fuel consumption in urban transport systems and encouraging new energy and new technology development and application. Its milestones targets are as follows: by 2010, complete assessment of the existing urban transportation system and design low-carbon traffic integration solutions, bring the increase of energy-intensive and highly polluting vehicles under control, and encourage the use of environmentally friendly vehicles, new energy vehicles and electric vehicles; by 2013, have 14 compressed natural gas stations constructed and increase the share of gas-fuel buses and taxis of all vehicles to more than 20%; by 2015, construct a rapid transit system, and build a fast and efficient public transportation network in the urban area and among the county centers.

6.3.4 Low-Carbon Environment

The air quality in Baoding showed notable improvement. Baoding's development of the industrial economy brought along air pollution because of the sulfur dioxide, soot and dust emissions from industrial activities, making negative impacts on citizens' lives and health. In the past few years, Baoding imposed stricter market access criteria for energy-intensive and highly polluting enterprises and closed down small thermal power and small cement production enterprises. During 2005–2010, the number of days with an API of less than 100 in Baoding increased from 84.38% to 90.68%, higher than the national average which rose from only 51.9% to 81.7%.

Water shortage poses a challenge to Baoding's sustainable development. Baoding's water resources per capita is far lower than the national average. Due to continuous overexploitation of groundwater, there is less and less available. Since there is no waste water treatment plant yet, water pollution poses further challenges to water resources. It is imperative for Baoding to carry out water conservation work in particular. As indicated earlier, China's per capita domestic water consumption decreased from 149.8 kg in 2005 to 104.04 kg in 2010; however, per capita domestic water consumption in Baoding increased from 71.05 kg in 2005 to 96.63 kg in 2010, showing a reverse trend. Considering Baoding's per capita domestic water consumption is still lower than that of the national average, such growth is seen as reasonable.

From 2005 to 2010, Baoding's forest coverage rate showed an increasing trend overall. As of the end of 2010, Baoding's forest coverage rate was 18.86%, an increase of 1.32% compared to the Eleventh Five-Year Plan period. However, such a rate is still rather low. In the past few years, Baoding launched afforestation projects to increase forest coverage, yet young forests tend to be single-species with low ecological capacity. Soil erosion remains a challenging problem.

6.3.5 Low-Carbon Society

Baoding's urban-rural income ratio during 2005–2010 was between 2.5 and 2.763, lower than the national average of 3.22 over the same period. This shows that the income gap in Baoding is smaller.

Baoding's CO_2 emission per capita increased from 3.219 tCO_2/person in 2005 to 4.312 tCO_2/person in 2010. In the same period, the national average went from 4.34 to 5.67. Baoding's per capita

CO_2 emission is lower than the national average but its growth rate is higher.

In terms of low-carbon management system, China has set explicit goals to achieve low-carbon transition. Pilot projects are conducted and the policy framework is being continuously improved. Baoding developed its strategy of becoming China's Power Valley in 2006 and formulated a series of low-carbon oriented policies, including Baoding People's Government's Guide for the Construction of Low-Carbon City (Trial), Baoding People's Government's Implementation Guide for the Construction of the Sun City and Guides of Baoding People's Government on the Construction of Low-Carbon Life. The last one has set goals, tasks, key projects and safeguarding measures for the coming ten years, becoming the outline and roadmap for Baoding's low-carbon transition. The solar projects implemented expand solar application to various fields in production and citizens' lives.

6.4 Low-Carbon City Construction and Management in Baoding

This section of the report gives an overview of Baoding low-carbon city construction and management based on LCCC supporting indicators and their action checklists from four areas: urban management, green economy, green building and low-carbon transport. Overall the assessment shows that Baoding has been actively promoting its low-carbon development and has made achievements.

6.4.1 *Urban Management*

By focusing on its low-carbon development goals, Baoding established an Energy Conservation Work Leading Group with clear tasks and responsibilities to guide the low-carbon work. Baoding conducts its low-carbon work through administration and governance, i.e. through integrating low-carbon concepts into its urban planning, urban construction and urban infrastructure development, public services and public activities.

First, Baoding issued a Baoding Low-Carbon City Construction Plan (2011–2020) ("Low-Carbon Plan" hereafter), setting explicit targets of a 35% reduction in the city's carbon dioxide emission per unit of GDP from the 2005 level by 2015, and by 48% in 2020 relative to the 2005 level. The plan also proposed supporting policies, including industrial policies and fiscal policies to encourage low-carbon development and energy conservation by traditional industries. Baoding organized a Low-Carbon Development Inception Meeting to publicize the Low-Carbon Plan and increase information disclosure to the relevant authorities and the public. Low-carbon capacity-building activities were conducted afterwards.

The Low-Carbon Plan emphasizes the development of local renewable energy resources. Based on local energy resources endowment, priorities are given to solar photovoltaics, biomass, geothermal, waste-to-energy projects, hydro generation and other renewable energy alternatives to partially replace conventional energy consumption. The share of renewable energy is aimed to be increased step by step. Baoding prepared the Baoding New Energy Urban Development Plan and set specific targets of new energy development: the total installed capacity of new energy including solar, wind, biomass, waste incineration and hydro is to reach 320 MW by 2015 and 600 MW by 2020.

Meanwhile, the Low-Carbon Plan requires traditional industries to accelerate their pace of achieving energy-saving goals and to identify priority industries for energy-saving projects according to their carbon intensity — electricity industry, gas industry, chemical fiber manufacturing industry, building materials industry and chemical industry. To ensure such energy-saving emission reduction policies are well implemented in these industries, Baoding established an expert team to start the preparation of greenhouse gas inventories and improve energy and carbon emissions database. At the same time, Baoding Finance Bureau issued the Management of Specific Energy Saving Budget, which set the target of increasing the low-carbon budget from 6 million *yuan* in 2010 to an increase of 10 million *yuan* in 2012.

Second, low-carbon concepts are gradually integrated into the planning and development of municipal facilities, including heating, water supply, waste disposal, etc. Take heating as an example. Baoding adjusted the temperature of the primary and secondary pipe network in the heating season according to the outdoor temperature conditions in order to achieve energy saving. Meanwhile, a heating online monitoring system has been established, with all new thermal stations equipped with a remote monitoring system for real-time monitoring and remote control. Piped natural gas coverage in households reached 99.52% by the end of 2010, serving 325,000 households. To promote water conservation, Baoding issued several policies including Baoding Planned Water and Water Conservation Interim Provisions, Baoding Water Management Interim Provisions and Implementation Plan for Creating a Water-Saving City. To enhance energy management in water supply, Baoding has developed Energy Management Regulations, Regulations on Energy Conservation in the Summer and other regulations. However, there is no water-associated online energy monitoring programs or waste-water-associated energy management policy tools introduced yet.

At present, although the waste disposal rate in Baoding has reached 100%, there are no specific regulatory agencies, policies or infrastructure available for waste separation and reduction. For this reason, Baoding has issued the Notice on Integrated Waste Management and Baoding Waste Management Park Construction Plan which set specific objectives: by 2015, the collection and transport rate will reach 100%, the sanitary treatment rate will reach 100% and the recycling and utilization rate will reach 60% in built-up urban areas; and such rates will further increase to 85%, 85% and 50%, respectively, in planned urban areas. Waste separation and reduction projects have been planned for built-up urban areas, including new large waste transfer stations with a capacity of over 1000 t, 20 large closed waste transfer vehicles, waste sorting and transfer stations, and a standardized close system for waste collection and transport.

Baoding particularly pays attention to the leading role of demonstrative projects. It developed a Low-Carbon Community Award Program in 2011 and selected 3 low-carbon hospitals, 20 low-carbon schools, 3 low-carbon supermarkets, 15 low-carbon communities and 1 low-carbon hotel. The low-carbon hotel pilot, the Power Valley

International Hotel, features BIPV and a sewage source heat pump system. The hotel is the country's first building with a solar photovoltaic façade which generates electricity that is directly sent to the national power grid, saving 105 tce every year. The sewage source heat pump system recycles heat from the sewage network with very high efficiency of heat recovery, resulting in energy savings of 40–60% compared to average air conditioning equipment.

In order to increase low-carbon awareness among the public, Baoding actively publicizes its low-carbon policies, regulations and projects through newspapers, radio, television, brochures, the Internet and other forms of mass media. It organized seminars and forums to encourage participation by the general public. Baoding participated in Earth Hour organized by the WWF over four consecutive years. Meanwhile, the government brought low-carbon knowledge into schools and communities to spread low-carbon concepts. It also called for public participation by asking them to choose green modes of travel at least once a week, reduce TV watching by one hour, hand wash their clothes at least once and ride the elevator once fewer.

6.4.2 *Green Economy*

The secondary sector of the economy accounts for a larger proportion of GDP in Baoding. Meanwhile Baoding is also a major agricultural city, the 22 counties and 5 districts under its jurisdiction have a much larger rural area than urban area. Therefore, this section mainly analyzes Baoding's low-carbon economy development through researching its policies on agriculture, forestry and industry while taking into account the tertiary sector.

Based on the principle of comprehensive utilization of resources, the Baoding government actively promoted low-carbon agriculture development with projects aiming at saving water, fertilizer, medicine, seeds, land and energy. Such projects include Ecological Homes, soil testing and fertilizer projects, comprehensive utilization of straw and agricultural water-saving projects. In 2011, Baoding had 445,000 biogas digesters and a biogas penetration rate reaching 20.2%. Annual biogas production was $13,400\,m^3$, saving energy of 95,600 tons of coal equivalent while generating economic benefit of 576 million *yuan* and benefiting a population of 1,770,000. The total number of soil samples tested was 129,900, based on which 38 formulations were designed, more than 300 fertilizer formulae were engineered and 27,031,000 *mu* of land applied new fertilizer formula that resulted in total avoided fertilizer use of 221,700 tons and generated 1,219,098,100 *yuan* of economic benefit. The technology of returning straw to land was applied in 9.4 million *mu* of land, and the straw utilization rate grew from 75% in 2006 to 86% in 2011. Improved irrigation practices were used in 4.40 million *mu* of land, resulting in water saving of 190 million m³ from wheat farms and 0.90 million m³ from vegetable farms.

Baoding combined low-carbon forestry development with park and forest development programs, issuing a number of policies with the goal of increasing carbon sinks and saving energy. Baoding planted 358,200 *mu* of forest and closed 340,000 *mu* of forest for protection. Among this 146,800 *mu* of new forest and 169,000 *mu* of closed forest were funded by a budget from higher level investment, while all the remaining 211,400 *mu* of new forest and 171,000 *mu* of closed forest were completed with local financial support, voluntary tree planting and donations. Baoding has made achievements in reforming forest property rights. Based on relevant national policies it identified clear property rights for the collective forest area of 12,250,000 *mu*, completing 99% of its forest property reform tasks. Ownership of 1,148 *mu* was confirmed and property certificates were issued, accounting for 93% of its reform tasks. One hundred and forty-four items of forest ownership trading were conducted, concerning forest of 330,000 *mu* and 43.21 million *yuan*. Eighty-eight forest mortgage loans were issued, concerning forest of 40,000 *mu* and total loan amount of 119.43 million *yuan*. Insurance was purchased for forest of 10,900 *mu* with a total amount of 43.39 million *yuan*.

To support the low-carbon transition of its industries, Baoding issued Further Strengthening Efforts to Promote the Implementation Scheme of Energy Saving and Emission Reduction for the Twelfth Five-Year Plan Period, Baoding 2012 Implementation Scheme of Energy Saving and Emission Reduction, and Notice on Allocating

2012 Tasks of Energy Saving and Emission Reduction. Such policy documents impose stricter market access criteria for energy-intensive and highly polluting enterprises, and help enterprises to get national and provincial energy-saving technological transfer and financial incentive funds. Baoding has developed plans to phase out small thermal power plants and small cement production enterprises, established an energy monitoring system for industrial enterprises and key energy users above a certain scale, promoted the application of energy management systems, and signed letters of energy conservation targets with key energy-consuming enterprises. Evaluation of the implementation of such agreements was carried out through reporting, field visits and document checking.

With the introduction of those above-mentioned policies, photovoltaics, wind power and other new energy low-carbon industries experienced rapid development in Baoding during the Eleventh Five-Year Plan period, and industries made notable achievements in energy saving. According to Baoding's Twelfth Five-Year Plan, the goal is that by 2015, the key new energy industry area reaches 100 km² with a total investment of 200 billion *yuan* and total turnover of 2,000 billion *yuan*. In terms of renewable energy, six industries were identified as priority ones for development: photovoltaics, wind power, electricity saving, energy storage, power transmission and power electronics, together forming industrial clusters in China's Electricity Valley and national solar demonstration zone. By 2015 the share of renewable energy is aimed to reach 6.5% in urban energy consumption in the pan-Baoding area.

Service industries have shown a good development trend in Baoding. Since 2010, Baoding has developed and implemented policies to encourage energy service companies to register and conduct business, and provided financial incentives for energy management contracts and CDM services. Meanwhile, two conferences are convened in Baoding annually featuring industrial enterprises to promote the concept of energy service. An active and regulated energy services market has been formed with both integrated energy service companies and specific-technology-based energy companies providing energy audit and energy assessment services. Training courses are offered to introduce energy management systems. Lecturers are from the local energy monitoring commission and low-carbon development research centers. Besides, regular experience-exchanging events are organized for public institutions regarding energy management. Documentation on the trainings is available.

6.4.3 *Green Building*

Green buildings refers to buildings which can save resources (energy, land, water and materials) to the maximum extent during their life cycle, that can help protect the environment and reduce pollution, provide people with healthy, convenient and efficient space, and can harmoniously coexist with nature. Development of green building is dependent on policy support.

Green building targets have already been incorporated into Baoding's master plan — establishing green building standards and technological system, and promoting green building projects. It is planned that the share of green buildings in total newly constructed buildings should reach 20% by 2015 and further to 80% by 2020. The government has actively introduced national policies including the MOHURD Twelfth Five-Year Plan of Green Buildings and Implementation Guides to Accelerate Green Building Development in China by the Ministry of Finance and MOHURD (MOF-MOHURD [167] to lower level government and developers in their daily work.

Baoding takes green building labeling seriously. It has organized seminars to increase awareness among decision-makers in the building sector every year. Special technical services and support is provided to key projects. Green building seminars are conducted one to two times every year to provide professional training. Meanwhile, focusing on saving land, energy, water and materials and achieving environmental protection, Baoding provides support to developers in applying national green building certification. Experts are hired from the Baoding Green Building Expert Pool and from China Architecture and Building Researchers to give guidance for the 14 projects' application materials. By the end of 2012, a total of six projects in Baoding received national green building certifications, while seven projects were chosen as Top Ten Green Buildings in Hebei or Top Ten Green Communities in Hebei.

Baoding attaches great attention to building energy consumption statistics. It started to build a database in 2009 in accordance with the Comprehensive Energy Consumption Statistics Table for Urban Civil Buildings. A closed management cycle is established covering energy-saving project evaluation, construction drawings review, process inspection, special inspection, etc., in accordance with the national policies including Building Energy Efficiency Evaluation and Labeling Technical Guidelines, Building Energy Efficiency Evaluation and Labeling Management Guidelines and other relevant regulations. An energy consumption quota is developed for buildings. It is planned that a data survey will be completed for office buildings and newly constructed large public buildings among its 27 counties. Meanwhile, Baoding has helped Hebei University, Hebei Agricultural University and Hebei Institute of Finance to establish campus energy monitoring platforms. An energy consumption monitoring platform for Baoding's buildings has also been built.

Baoding actively promoted renewable energy application in buildings. To strengthen the management, it took the lead in Hebei Province in implementing a filing system and closed regulatory cycle for renewable energy applications. For buildings that should apply renewable energy but actually did not do so, they will fail the filing and acceptance. To promote the construction of demonstration projects with renewable energy applications, Baoding issued four supporting policies including Implementation Guides on Constructing Renewable Energy Demonstration City in Baoding and Baoding Management Guides on Demonstration Building Project with Renewable Energy Applications to support demonstration building projects with solar photovoltaics. Currently, there are 3.6 million m² of construction projects applied with renewable energy, and seven national demonstration buildings integrated with photovoltaics were checked and accepted.

Energy-saving retrofitting of existing buildings received government support as well. On the one hand, Baoding has carried out thorough surveys on existing buildings and identified goals, tasks and measures for retrofitting purposes. On the other hand, it has guided subsidiary counties such as Mancheng, Wangdu, Anxin, Laishui and Xushui in carrying out energy-saving retrofits for buildings. In addition, it has implemented the ban on solid concrete bricks. First, from the source the city closed down 29 solid concrete brick manufacturing enterprises with a total production capacity of 560 million standard bricks and retrieved land of 2,996 mu. Then it launched 28 enterprises with new wall materials production lines. The total amount of new wall materials produced accounted for more than 80% of all materials. Up to now, Baoding has completed energy-efficient buildings of over 10 million m², which can achieve energy savings of 120,000 tons of coal equivalent and avoid 300,000 tons of carbon dioxide emissions.

6.4.4 *Low-Carbon Transport*

The transportation system is an integral part of urban construction and urban modernization. Low-carbon transport refers to a transport system featuring high energy efficiency, low energy consumption, low pollution and low emissions. It can be achieved through increasing energy efficiency, upgrading the energy mix and improving development models. Baoding explored the possibilities of building a low-carbon transport system according to its local conditions.

Baoding has developed the Implementation Scheme for Constructing Low-Carbon Transport Pilot in Baoding and Implementation Scheme for Constructing Green and Low-Carbon Transport Regional Pilot Project (2013–2020), which set goals in each phase. As Baoding was identified as one of the first ten low-carbon transport system pilot cities by the Ministry of Transport, it also carried out exploratory work to achieve energy saving in the transport sector including building green corridors along roads, increasing the gasification rate, applying green lighting technologies for street lighting and building an intelligent transport system step by step.

Baoding implemented the strategy of increasing the gasification rate. First, it developed the Implementation Scheme for Using

Liquefied Natural Gas for Buses and Long-Distance Passenger Vehicles to promote clean energy use in the public transport system. In 2012 the government purchased 689 units of natural gas buses to replace all non-gas buses. Meanwhile, it developed a plan to replace the current fleet with 1,170 LNG buses. By the end of 2012 a total of 260 LNG vehicles were put into operation. Second, it implemented dual-fuel taxi transformation. Among the city's 3,036 taxis 2,500 completed the oil to gas transformation, accounting for 82.3% of the total number of taxis, reducing carbon dioxide emissions of 15,600 kg per day and saving 12,813 tons of coal equivalent. Furthermore, Baoding started constructing urban natural gas recharge stations, with six completed and four under construction.

Green lighting technologies were applied. First, a series of measures were taken to reduce street lighting energy use, including strictly controlled switch time, using indirect lighting technologies at midnight, installing electricity-saving devices, applying LED lights and raising design standards. Second, LED lights were installed for tunnels along two highways, which in estimation will lead to energy savings of 771 tce and carbon dioxide emission reduction of 2,052 tons annually. Third, LED lights were installed in eight service areas and parking lots along the Beijing–Kunming highway, which in estimation will lead to electricity savings of 10,732,900 kWh (equal to 3,413.06 tce) and carbon dioxide emission reduction of 8,324.45 tons annually.

Baoding has made some achievements in its intelligent transport system construction. First, it improved intelligent scheduling platform for taxis. All 2,428 urban taxis are installed with GPS systems and talk back equipment, enabling integrated scheduling and telephone reservation that will help reduce the empty-loaded rate and increase productivity. Second, monitoring platforms were upgraded for national and provincial trunk highways. Currently, the monitoring system covers 60% of national and provincial roads in Baoding and two high-speed highway traffic monitoring command centers have been established. Third, key commercial vehicles were installed with GPS monitoring devices. Currently, 80% of all passenger and 100% of dangerous goods transport vehicles are monitored to reduce the empty-loaded rate.

6.5 Challenges

From the LCCC Primary Indicator System evaluation results, it can be concluded that overall Baoding has made good achievements in its low-carbon city construction. It performed not so well in some indicators due to its resource endowments. For example, Baoding's forest coverage rate is lower than the national average level, its carbon productivity is much higher than the national average level, its energy mix is coal-dominated and the share of non-commercial renewable energy use is very low. Despite efforts to improve, it is still lagging behind in these aspects.

Hebei Province proposed the strategy of developing a Pan-Beijing Green Economy Belt with 14 counties, 1 circle, 4 districts and 6 bases in order to safeguard the resource supply to Beijing and increase the region's competitiveness. Located in a province with rapid economic restructuring as Hebei, Baoding enjoys the geographical advantage of being next to Beijing and Tianjin, has a convenient and three-dimensional transport system and owns rich natural resources and abundant intellectual support, yet its economic development is relatively slow compared with other prefecture-level cities in the province. The disposable income of urban residents in Baoding is rather low compared to other cities in the province.

Figure 6.2 shows the GDP of all prefecture-level cities in Hebei Province in 2010. Although Baoding ranks fifth among the eleven prefecture-level cities, its GDP is less than half of that of Tangshan which ranks number one. In 2010, Baoding's urban residents per capita disposable income was 15,047.68 *yuan*, 4,061.76 *yuan* lower than the national average of 19,109.44 *yuan* in the same period, also lower than the provincial average of 16,263.43 *yuan* in Hebei. Urban residents' per capita consumption expenditure was 9,625.75 *yuan*, 3,845.7 *yuan* lower than the national average. Rural per capita net income was 5,446 *yuan*, 473 *yuan* lower than the national average of 5,919 *yuan* (see Table 6.2) and also lower than the provincial average of 5,957.98 *yuan*.

The Twelfth Five-Year Plan period is a critical phase for Baoding to transition from a traditional energy-intensive pattern to a

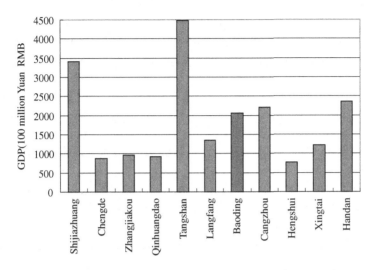

Figure 6.2 GDP of Prefecture-Level Cities in Hebei (2010).

Data source: Hebei Yearbook 2011.

Table 6.2   Economic Indicators of Prefecture-Level Cities in Hebei (2010).

	Urban Residents		Rural Residents	
	Disposable Income	Consumption Expenditure	Net Income	Domestic Consumption
Shijiazhuang	18,289.95	10,568.49	6,577	3,956
Chengde	14,667.55	9,490.05	4,382	3,672
Zhangjiakou	14,649.46	9,873.63	4,119	3,110
Qinhuangdao	17,202.83	11,081.06	6,214	4,070
Tangshan	19,555.66	13,521.96	8,310	5,980
Langfang	19,576.35	12,672.77	7,589	3,850
Baoding	15,047.68	9,625.75	5,446	2,864
Cangzhou	16,115.96	10,278.99	5,528	3,525
Hengshui	14,525.23	9,211.29	4,370	2,779
Xingtai	14,744.43	10,416.34	4,966	2,864
Handan	17,561.62	9,437.82	6,085	2,692

Note: Data above are per capita value, with a unit of RMB *yuan*.

Data source: Hebei Yearbook 2011.

sustainable low-carbon development pattern. Baoding has a strong urge to pursue economic development and catch up with peer cities, and such an urge makes it necessary for Baoding to adhere to its strategy of low-carbon development and explore a roadmap based on its actualities.

Urban low-carbon management is a comprehensive, complex, systematic process that requires proper match policy instruments as support. Although the Guides of Baoding People's Government on Building Low-Carbon Life (Trial) has set goals, tasks, key projects and safeguard measures for its low-carbon development, there are a lot more heavy tasks to be implemented to achieve success, including constructing an energy supervision system and developing fiscal incentives. First, improvement is needed for emissions monitoring, statistical systems, feedback and long-term monitoring mechanisms. In particular, the city's overall carbon emissions inventory needs to be developed and energy management requirements need to be set for waste water treatment. Second, the design and implementation of fiscal tools need to be accelerated. Currently, Baoding has not established a mechanism to evaluate the usage of low-carbon development funding; hence there is no good way to regularly check and assess whether the fund usage meets original expectations. Moreover, the Baoding government has not set green procurement requirements in its plan. In fact, the city has not yet prepared the list of green products or other necessary policies. Third, a low-carbon information disclosure platform needs to be established to inform and involve the public.

With the implementation of supporting policies, photovoltaics, wind power and other new energy low-carbon industries in Baoding experienced rapid development, while industrial energy saving achieved remarkable results. However, the economic policies did not change the carbon-intensive nature of its industrial structure, and the growth of energy-intensive industries is still rapid. First, Baoding has large room for improvement in terms of energy management, such as industrial enterprise energy monitoring system construction, elimination of obsolete production capacity, special DSM fund usage, control of non-CO_2 greenhouse gas emission in industrial processes, as well as promotion of voluntary emissions reduction agreements. Second, the overall industrial structure needs to be further upgraded. A mature industrial structure has both leading industries that take regional comparative advantages and supporting industries, together forming interactive industrial clusters. Currently the new energy industry is apparently the leading industry in Baoding, yet it is to be studied how to strengthen the cooperation between new energy industries and its other pillar industries such as automobile production. Finally, the new energy industry in Baoding already has a good foundation and advantages, but it still faces the bottleneck of intertwined challenges in capital, talent, technology and product marketing.

In the low-carbon development residents in Baoding played an invaluable role. In order to increase the public awareness of low-carbon development, Baoding actively advocates its visions and policies through newspapers, radio, television, brochures, the Internet and other forms of mass media. Such efforts turned out to be effective. A survey shows that the level of difficulty in implementing a policy is correlated with the convenience it brings to the people. Building and transportation projects were particularly difficult to implement. Retrofitting of existing buildings and low-carbon district construction in new areas are two priorities for future work in Baoding. To further promote the integration of solar facilities in buildings, the government integrated building retrofits and renewable energy in its architectural plan and master plan, provided financial support for building retrofit and solar energy application, and conducted seminars and workshops. With rapid urbanization, it becomes imperative to compose a plan, action list and implementation guides for new construction energy management. In the transport sector, the smooth progress of the gasification strategy is gradually changing the energy consumption patterns in the public transport sector. It is strategically important to improve the city's slow traffic system both to support eco-city construction and to benefit the growing elder generation.

6.6 Conclusion and Recommendations

Baoding has both highlights and shortcomings in its low-carbon city construction. It is remarkable how the city developed its solar energy industry clusters and raised overall awareness of sustainable development among the general public. Its disadvantages include a carbon-intensive energy mix, energy-intensive industrial structure, lack of an integrated plan and concrete measures for low-carbon development.

First, Baoding needs to compile a low-carbon inventory and develop low-carbon city management tools as the basis for policy-making. According to the LCCC indicator system recommendations, the tool should include three databases or software: first, the urban carbon emission inventory which is a top-down model established with statistics collected and used for monitoring and forecasting the trend of different sectors; second, energy consumption data on different types of buildings; third, an emission inventory of different transportation modes including buses, private cars, taxis and so on. Take transportation for example. Baoding has linked vehicle exhaust testing with vehicle annual checks to strengthen regulation on gas emissions and encourage the use of new energy vehicles.

Second, Baoding could improve the implementation of policy incentives on new energy investment and consumption. The government should establish good feedback channels and enforce periodic verification of the use of funds, comparing it with original expectations. In order to encourage investment and consumption of renewable energy, government procurement should be utilized to leverage the market. The first priority is to develop a green product list consisting of energy-saving, water-efficient, low pollution, low toxicity, renewable and recyclable products with reference to the national list. Meanwhile, a green procurement plan, rules and requirements should also be developed.

Third, it is necessary to establish a technical information dissemination platform to promote low-carbon technologies among farmers. It is also recommended to integrate social research resources and financial resources and start scaling up programs of low-carbon technologies. At present, low-carbon technologies are disseminated mainly through a family-to-family approach. Technical specifications and guidance should be developed focusing on treatment of agricultural and forestry residues and agricultural energy conservation. More integrated scaling-up models should be considered. The rich intellectual resources in universities in Baoding can be utilized and multi-channel funding channels can be explored to support such scaling up.

Finally, Baoding should continue to promote low-carbon projects in municipal infrastructures, transportation and buildings while building an interactive information platform to guide lifestyle changes among local residents. Propaganda activities should be designed for different audiences and proper media should be chosen accordingly. Moreover, while developing low-carbon policies the government should fully consider the people's interests, so that the future implementation of such policies will increase social welfare. In addition, the establishment of an information platform through radio, newspapers and other media can provide the public with a channel to give feedback to low-carbon plans, policies and projects.

References

1. Liu Qian. 2012. The position and development of Baoding in Hebei pan-capital-city. *Green Economy Circle*, 20: 153–154.
2. Gao Yumin, Lu Jifeng, Liu Dazhong. 2012. Hebei pan-capital-city green economy circle integration coordination mechanism. *Journal of Guizhou Ethnic Institute (Philosophy and Social Sciences Version)*, 132(2): 162–164.
3. Zhang Zuxin, Li Wei. 2012. Industrial carbon emission in Baoding. *Business Culture*, 4: 315.
4. Wei Yong. 2010. Low-carbon Baoding, ecological ancient city — Low-carbon city construction report. *Statistics and Management*, 5: 64–65.
5. Li Jianming, Wen Jing. 2011. Low-carbon city development roadmap — Case study on Baoding City in Hebei Province. *Theorie*, 454(7): 34–36.
6. Baoding Municipality. 2011. *Baoding Low-Carbon City Construction Plan (2011–2020)*.
7. Baoding Development and Reform Commission. 2010. *Baoding New Energy City Development Plan*.
8. Baoding Financial Bureau. 2009. *Management of Budget for Energy Efficiency and Emission Reduction*.
9. Baoding Municipality. 2010. *Baoding Water Usage Plan and Water Conservation Regulations (Interim)*.
10. Baoding Municipality. 2010. *Baoding Water Resource Management Guidelines (Interim)*.
11. Baoding Municipality. 2010. *Implementation Plan of Building Water Efficient City of Baoding*.

12. Baoding Water Supply Co. 2010. *Energy Management Regulations.*
13. Baoding Water Supply Co. 2010. *Notice on Energy Saving in Summer.*
14. Baoding Urban management Law Enforcement Bureau. 2010. *Notice on Integrated Management of Municipal Solid Waste in Baoding.*
15. Baoding Urban management Law Enforcement Bureau. 2010. *Construction Plan of Baoding Waste Treatment District.*
16. Baoding Municipality. 2010. *Low-Carbon Community Selection Criteria.*
17. Baoding Municipality. 2011. *Implementation Guide on Further Promoting Energy Efficiency and Emission Reduction during 12th Five-Year Plan Period.*
18. Baoding Municipality. 2011. *Baoding Implementation Plan on Energy Efficiency and Emission Reduction 2012.*
19. Baoding Municipality. 2011. *Notice on Allocating 2012 Energy Efficiency and Emission Reduction Targets.*
20. Baoding Municipality. 2011. *Special Plan on Baoding Renewable Energy Application in Buildings* (2011–2015).
21. Baoding Municipality. 2011. *Implementation Plan on Renewable Energy Application in Buildings* (2011–2013).
22. Baoding Municipality. 2011. *Renewable Energy Application in Buildings Pilot City Program — Implementation Guide.*
23. Baoding Municipality. 2011. *Renewable Energy Application in Buildings Pilot City Program — Management Guidelines.*
24. Baoding Municipality. 2011. *Implementation Guide to Promote Renewable Energy Application in Buildings.*
25. Baoding Municipality. 2011. *Renewable Energy Application in Buildings Pilot City Program — Implementation Guide.*
26. Baoding Municipality. 2011. *Implementation Plan on Constructing Low-carbon Transport Pilot Projects.*
27. Baoding Municipality, Transport Science Research Institute. 2011. *Implementation Plan on Constructing Low-carbon Transport Regional Pilot Projects.*
28. Baoding Municipality. 2009. *Interim Guidelines on Building Low-carbon Lifestyles.*

Appendix: Recommended Low-Carbon Action Plan for Baoding

#	Recommended Action	Details	Dept in Charge	Priority	Deadlines (year)
I.	***Urban management***				
1.	Preparation of comprehensive carbon emissions inventory of the city	Establish emission monitoring, statistical, feedback and long-term regulatory mechanisms. Compile emissions inventory based on statistics system	DRC	High	2
2.	Establish evaluation mechanisms for the effect of funds allocated	Establish feedback channels and periodic verification of funds, compare with expected results	FB leading, others supporting	Low	1
3.	Integrate green procurement into government planning	Develop Green Product Directory based on green procurement list to regulate procurement	Government Purchase Office	High	1
4.	Prepare green products directory and supporting files	Prepare the directory based on principles such as energy/water saving, low pollution, low toxicity, renewable and recyclable with reference to the country's list	Government Purchase Office	High	1
5.	Establish information platform for transparency of low-carbon planning and management information	Establish information platform via the Internet, radio, newspapers, etc., to publicize low-carbon planning and management information	DRC	High	2

(Continued)

(*Continued*)

#	Recommended Action	Details	Dept in Charge	Priority	Deadlines (year)
6.	Build channels for feedback on low-carbon planning and management	Through newspapers, administrative agencies and other communication channels	DRC	Medium	1
7.	Collect comprehensive data on renewable energy production	Collect comprehensive data on renewable energy production	DRC, HURD, SB	High	1.5
8.	Set clear water-associated energy management targets	Set feasible targets through surveys and discussions	WAB	Medium	0.5
9.	Set clear regulations for sewage treatment plant energy management	Responsible department issues explicit regulation guidelines in accordance with local actualities	PUB — Sewage Co.	High	1
10.	Set clear waste-water-associated energy management targets	Set feasible targets and integrate them into urban planning	PUB — Sewage Co.	High	1
11.	Develop water conservation action plan	Responsible departments set targets and action plans respectively for residential and industrial water use reduction	WAB — Water Saving Division	Medium	1
12.	Develop waste treatment pilot projects and sector standards	Develop waste treatment pilot projects and sector standards considering different phases: collection, sorting, transfer and treatment	EMLEB	Low	2

(*Continued*)

(*Continued*)

#	Recommended Action	Details	Dept in Charge	Priority	Deadlines (year)
13.	Increase awareness of waste reduction and recycling	Increase awareness of waste reduction and recycling through propaganda events and pilot projects	EMLEB	Medium	1
II. Green economy					
14.	Get GHG emission statistics for industrial processes	Establish mechanisms for emission monitoring, statistics collection, feedback and long-term regulation of industrial enterprises. Build emissions inventory accordingly	IIC, SB and DRC	High	2
15.	Develop systematic DSM planning	Develop systematic DSM planning from consumption perspective	DRC (Energy Division)	High	1
16.	Promptly publicize energy monitoring results	Promptly publicize energy monitoring results	IIC	Medium	0.5
17.	Set specific targets for renewable energy use among industries	Set specific targets for renewable energy use among industries based on urban master plan	DRC, IIC	High	1
18.	Regularly write and publish reports on industrial solid waste utilization	Set special projects and funds to write and publish reports on industrial solid waste utilization	EPB, IIC	Medium	1

(*Continued*)

(*Continued*)

#	Recommended Action	Details	Dept in Charge	Priority	Deadlines (year)
19.	Appoint full-time energy manager with clear responsibilities in public institutions	Leading institution should supervise that at each responsible departments, an full-time energy manager with clear responsibilities is appointed	Public Institution Office, CB, HURD	High	1
20.	Sign voluntary emissions reduction agreements with service companies to promote low-carbon consumption	Develop template, conduct publicity campaign and encourage service companies to sign voluntary emissions reduction agreement	CB	High	2
21.	Provide policy incentives to encourage energy service companies to register and operate	Provide (both material and non-material) incentives to encourage energy service companies to register and operate	DRC, IIC	Medium	1
22.	Provide relevant information and technical support for producers regarding agriculture and forestry residue treatment and reuse	Provide relevant information (including manuals, notices and documents) and technical support for producers regarding agriculture and forestry residue treatment and reuse	AB, Forest Bureau	High	2

(*Continued*)

(*Continued*)

#	Recommended Action	Details	Dept in Charge	Priority	Deadlines (year)
23.	Issue policies that encourage farmers to get certified for green/organic/pollution-free products	Issue policies that encourage farmers to get certified for green/organic/pollution-free products	AB	High	1
24.	Develop a plan to achieve agricultural energy savings	Responsible department develops a plan with specific measures	AB	High	2
III. Green building					
25.	Develop a sound work plan to collect building energy statistics (including the statistical classification for all building types)	Establish mechanisms for emission monitoring, statistical, feedback and long-term regulation. Compile emissions inventory based on statistics	SB, HURD	High	3
26.	Prepare energy use plan and action guidelines for new buildings in new urban area	Prepare energy use plan and action guidelines for new buildings in new urban area	HURD	High	1
27.	Develop mid/long-term plan for existing building retrofit	Develop mid/long-term plan, action plan and implementation guides for existing building retrofit	HURD	High	2

(*Continued*)

(*Continued*)

#	Recommended Action	Details	Dept in Charge	Priority	Deadlines (year)
28.	Preparation of new urban planning, new buildings, energy use, planning, action guidelines	Further promote building integration retrofit for solar and other renewable energy. Organize campaigns and provide policy and financial incentives	HURD	High	3
29.	Set RE application targets and control during drawing inspections	Strengthen management on RE application through process control during drawing inspections	HURD	High	2
IV.	*Low-carbon transport*				
30.	Start preparing carbon emissions inventory for the transport sector	Establish emission monitoring, statistical, feedback and long-term regulatory mechanisms. Compile emissions inventory based on statistics system.	TB (Integrated Planning Division)	High	2
31.	Establish emission statistics monitoring system for vehicles in operation	Collect data on vehicle types and efficiency, and develop emission statistics monitoring system for vehicles in operation	TB	High	2
32.	Smart transport construction	Develop effective bus dispatch center and optimize smart taxi dispatch center	TB	High	1

(*Continued*)

(*Continued*)

#	Recommended Action	Details	Dept in Charge	Priority	Deadlines (year)
33.	Clarify funding sources and the responsible departments to implement integrated transport	Clarify funding sources and the responsible departments to implement integrated transport	TB (Integrated Planning Division)	High	0.5
34.	Train procurement staff on new energy vehicles	Conduct regular training activities	TB (Policy and Regulation Division)	Medium	2
35.	Collect data on the number of existing road lamps and their energy consumption	Collect data on the number of existing road lamps and their energy consumption	PUB — Road Lighting Division	High	1
36.	Formulate guidelines and action plan that gives priority to public transport	Responsible departments take the lead to formulate guidelines and action plan that gives priority to public transport	TB, Bus Company	Medium	1
37.	Set specific measures to increase public transport service, including punctuality, convenience, comfort level and affordability	Conduct surveys and expert seminars to develop such specific measures	TB, Bus Company	High	1

(*Continued*)

(*Continued*)

#	Recommended Action	Details	Dept in Charge	Priority	Deadlines (year)
38.	Develop integrated plan for slow traffic key infrastructures	Develop integrated plan for slow traffic key infrastructures	Planning Bureau	High	1

Abbreviations

DRC: Development and Reform Commission

HURD: Bureau of Housing, Urban and Rural Development

IIC: Industrial and Information Commission

FB: Financial Bureau

PB: Planning Bureau

AB: Agriculture Bureau

CB: Commercial Bureau

TB: Transport Bureau

SB: Statistics Bureau

WAB: Water Affairs Bureau

PUB: Public Utilities Bureau

BERO: Building Envelop Retrofit Office

BEEST: Building Energy Efficiency Science and Technology Department, Public Building Energy Management Agency

TMDSTD: Traffic Management Division Safety and Technology Department

UMLEB: Urban Management Law Enforcement Bureau

Index

Printed in the United States
By Bookmasters